CHEMICAL ENGINEERING

An Introduction

"Chemical engineering is the field of applied science that employs physical, chemical, and biochemical rate processes for the betterment of humanity." This opening sentence of Chapter 1 is the underlying paradigm of chemical engineering. *Chemical Engineering: An Introduction* is designed to enable the student to explore the activities in which a modern chemical engineer is involved by focusing on mass and energy balances in liquid-phase processes. Applications explored include the design of a feedback level controller, membrane separation, hemodialysis, optimal design of a process with chemical reaction and separation, washout in a bioreactor, kinetic and mass transfer limits in a two-phase reactor, and the use of a membrane reactor to overcome equilibrium limits on conversion. Mathematics is employed as a language at the most elementary level. Professor Morton M. Denn incorporates design meaningfully; the design and analysis problems are realistic in format and scope. Students using this text will appreciate why they need the courses that follow in the core curriculum.

Morton M. Denn is the Albert Einstein Professor of Science and Engineering and Director of the Benjamin Levich Institute for Physico-Chemical Hydrodynamics at the City College of New York, CUNY. Prior to joining CCNY in 1999, he was Professor of Chemical Engineering at the University of California, Berkeley, where he served as Department Chair, as well as Program Leader for Polymers and Head of Materials Chemistry in the Materials Sciences Division of the Lawrence Berkeley National Laboratory. He previously taught chemical engineering at the University of Delaware, where he was the Allan P. Colburn Professor. Professor Denn was Editor of the *AIChE Journal* from 1985 to 1991 and Editor of the *Journal of Rheology* from 1995 to 2005. He is the recipient of a Guggenheim Fellowship; a Fulbright Lectureship; the Professional Progress, William H. Walker, Warren K. Lewis, Institute Lectureship, and Founders Awards of the American Institute of Chemical Engineers; the Chemical Engineering Lectureship of the American Society for Engineering Education; and the Bingham Medal and Distinguished Service Awards of the Society of Rheology. He is a member of the National Academy of Engineering and the American Academy of Arts and Sciences, and he received an honorary DSc from the University of Minnesota. His previous books are *Optimization by Variational Methods*; *Introduction to Chemical Engineering Analysis*, coauthored with T. W. Fraser Russell; *Stability of Reaction and Transport Processes*; *Process Fluid Mechanics*; *Process Modeling*; and *Polymer Melt Processing: Foundations in Fluid Mechanics and Heat Transfer*.

Cambridge Series in Chemical Engineering

Chemical Engineering

AN INTRODUCTION

Morton M. Denn
The City College of New York

CAMBRIDGE UNIVERSITY PRESS
Cambridge, New York, Melbourne, Madrid, Cape Town,
Singapore, São Paulo, Delhi, Tokyo, Mexico City

Cambridge University Press
32 Avenue of the Americas, New York, NY 10013-2473, USA

www.cambridge.org
Information on this title: www.cambridge.org/9781107669376

First published 2012

Printed in the United States of America

A catalog record for this publication is available from the British Library.

Library of Congress Cataloging in Publication data

Denn, Morton M., 1939–
Chemical engineering : an introduction / Morton Denn.
    p.   cm. – (Cambridge series in chemical engineering)
Includes bibliographical references and index.
ISBN 978-1-107-01189-2 (hardback) – ISBN 978-1-107-66937-6 (pbk.)
1. Chemical engineering.   I. Title.
TP155.D359   2011
660–dc22   2011012921

ISBN 978-1-107-01189-2 Hardback
ISBN 978-1-107-66937-6 Paperback

Contents

Preface

"Chemical engineering is the field of applied science that employs physical, chemical, and biochemical rate processes for the betterment of humanity." This opening sentence of Chapter 1 has been the underlying paradigm of chemical engineering for at least a century, through the development of modern chemical and petrochemical, biochemical, and materials processing, and into the twenty-first century as chemical engineers have applied their skills to fundamental problems in pharmaceuticals, medical devices and drug-delivery systems, semiconductor manufacturing, nanoscale technology, renewable energy, environmental control, and so on. The role of the introductory course in chemical engineering is to develop a framework that enables the student to move effortlessly from basic science and mathematics courses into the engineering science and technology courses that form the core of a professional chemical engineering education, as well as to provide the student with a comprehensive overview of the scope and practice of the profession. An effective introductory course should therefore be constructed around the utilization of rate processes in a context that relates to actual practice.

Chemical engineering as an academic discipline has always suffered from the fact that the things that chemical engineers do as professionals are not easily demonstrated in a way that conveys understanding to the general public, or even to engineering students who are just starting to pursue their technical courses. (Every secondary school student can relate to robots, bridges, computers, or heart-lung machines, but how do you easily convey the beauty and societal importance of an optimally designed pharmaceutical process or the exponential cost of improved separation?) The traditional introductory course in chemical engineering has usually been called something like "Material and Energy Balances," and the course has typically focused on flowsheet analysis, overall mass balance and equilibrium calculations, and process applications of thermochemistry. Such courses rarely explore the scope of the truly challenging and interesting problems that occupy today's chemical engineers.

I have taken a very different approach in this text. My goal is to enable the student to explore a broad range of activities in which a modern chemical engineer might be involved, which I do by focusing on liquid-phase processes. Thus, the student addresses such problems as the design of a feedback level controller, membrane separation and hemodialysis, optimal design of a process with chemical reaction and separation, washout in a bioreactor, kinetic and mass transfer limits in a two-phase reactor, and the use of a membrane reactor to overcome equilibrium limits on conversion. Mathematics is employed as a language, but the mathematics is at the most elementary level and serves to reinforce what the student has studied during the first university year; nothing more than a first course in calculus is required, together with some elementary chemistry. Yet we are able to incorporate design meaningfully into the very first course of the chemical engineering curriculum; the design and analysis problems, although simplified, are realistic in format and scope. Few students of my generation and those that followed had any concept of the scope of chemical engineering practice prior to their senior year (and perhaps not even then). Students enrolled in a course using this text will understand what they can expect to do as chemical engineering graduates, and they will appreciate why they need the courses that follow in the core curriculum.

There is more material in the text than can reasonably be covered in one semester. The organization is such that mass and energy balances can be given equal weight in a one-semester course if the instructor so desires. I prefer to emphasize the use of mass balances in order to broaden the scope of meaningful design issues; any negative consequences of deemphasizing thermochemistry in the introductory course, should the instructor choose to do so, are minimal. Much of what once formed the core of the traditional material and energy balances course is now covered in general chemistry, sometimes in a high school setting, and thermodynamics offerings in many chemical engineering departments have become more focused, with more emphasis on chemical thermodynamics than in the past.

Chemical Engineering: An Introduction incorporates material from an earlier textbook, *Introduction to Chemical Engineering Analysis* (1972), which Fraser Russell and I coauthored. I have added a great deal of new material, however, and removed a great deal as well. Much of what remains has been rewritten. Thus, this is not a new edition, but rather a new creation, with an important family resemblance to an earlier generation.

My PhD advisor was the late Rutherford Aris, whose insightful scholarship was matched by his strong commitment to education, which is reflected in his outstanding textbooks and monographs. Aris believed that students learn best when a subject is presented with rigor, and he wrote with a clarity and elegance that made the rigor accessible to everyone. I think that "Gus" would have approved of the approach presented in this textbook, even if his literary standards are unattainable, and I respectfully dedicate *Chemical Engineering: An Introduction* to his memory.

I am grateful to my colleagues at the City College of New York (CCNY), especially Raymond Tu and Alexander Couzis, for their encouragement and their willingness to use the evolving draft in the classroom, and I appreciate the willingness of the CCNY second-year students to work with us. I am, of course, grateful to Fraser Russell for his insights during our long collaboration and for his generosity in permitting me to use some of the fruits of our joint work. Peter Gordon of

Cambridge University Press enthusiastically supported this project, and Kim Dylla graciously permitted us to use her art on the cover. Finally, I am grateful to my colleagues at the Casali Institute of Applied Chemistry of the Hebrew University of Jerusalem, especially Gad Marom and Shlomo Magdassi, and to the Lady Davis Fellowship Trust, for hospitality and support while I was composing the final chapters of the book. My wife Vivienne's hand is hidden, but it is present throughout.

New York
February 2011

1 Chemical Engineering

1.1 Introduction

Chemical engineering is the field of applied science that employs physical, chemical, and biochemical rate processes for the betterment of humanity. This is a sweeping statement, and it contains two essential concepts: *rate processes* and *betterment of humanity*. The second is straightforward and is at the heart of all engineering. The engineer designs processes and tangible objects that meet the real or perceived needs of the populace. Some civil engineers design bridges. Some mechanical engineers design engines. Some electrical engineers design power systems. The popular perception of the chemical engineer is someone who designs and operates processes for the production of chemicals and petrochemicals. This is an historically accurate (if incomplete) image, but it describes only a small fraction of the chemical engineers of the early twenty-first century.

> Chemical engineering is the field of applied science that employs physical, chemical, and biochemical rate processes for the betterment of humanity.

Let us turn first to the concept of rate processes, which is the defining paradigm of chemical engineering, and consider an example. Everyone is familiar with the notion that medication taken orally must pass through the digestive system and across membranes into the bloodstream, after which it must be transported to the relevant location in the body (a tumor, a bacterial infection, etc.) where it binds to a receptor or reacts chemically. The residual medication is transported to an organ, where it is metabolized, and the metabolic products are transported across still more membranes and excreted from the body, perhaps in the urine. Each of these processes takes time, and the rate of each step plays an important role in determining the efficacy of the medication. Chemical engineers are concerned with all natural and man-made processes in which physicochemical processes that are governed by the rates at which the physical transport of mass, momentum, and energy and the chemical and biochemical transformation of molecular species occur. The example of the fate of medication, and the logical extension to devising procedures that optimize the delivery of the drug to the active site, is an example of *pharmacokinetics*, which has been an area of chemical engineering practice since the 1960s and has led to many important advances. In the sections that follow we will briefly examine this and other areas in which the chemical engineer's interest in rate processes has resulted in significant societal benefit. We do this to illustrate the applications to which the material covered in the remainder of this introductory text and the courses that follow can be applied, although our scope of applications will be far more limited.

1.2 The Historical Chemical Engineer

Chemical engineering began as a distinct profession at the start of the twentieth century, although elements of what are now considered to be core chemical engineering have existed for centuries and more (fermentation, for example, is mentioned in the Bible and in Homer). The discipline began as something of an amalgam, combining chemistry having an industrial focus with the mechanical design of equipment. The early triumphs, which defined the profession in the public eye, had to do with large-scale production of essential chemicals. The invention of the fluid catalytic cracking (FCC) process by Warren K. Lewis and Edward R. Gilliland in the late 1930s was one such advance. A fluidized bed is a column in which a rising gas carries particles upward at the same average rate at which they fall under the influence of gravity, producing a particulate suspension in which the particles move about rapidly because of the turbulence of the gas stream. Crude oil contacts a granular catalyst in the FCC and is converted to a variety of low-molecular-weight organic chemicals (ethylene, propylene, etc.) that can be used for feedstocks and fuel. The cracking reactions are endothermic (i.e., heat must be added). Residual carbon forms on the catalyst during the cracking reaction, reducing its efficiency; this carbon is removed by combustion in an interconnected reactor, and the exothermic combustion reaction produces the thermal energy necessary to carry out the endothermic cracking reactions. The process is very energy efficient; its invention was crucial to the production of high-octane aviation gasoline during World War II, and it is still the centerpiece of the modern petroleum refinery.

As noted previously, fermentation processes have existed throughout human history. The first industrial-scale fermentation process (other than alcoholic beverages) seems to have been the production of acetone and butanol through the anaerobic fermentation of corn by the organism *Clostridium acetobutylicum*, a conversion discovered in 1915 by the British chemist Chaim Weizmann, who later became the first President of the State of Israel. The production of acetone by this route was essential to the British war effort in World War I because acetone was required as a solvent for nitrocellulose in the production of smokeless powder, and calcium acetate, from which acetone was normally produced, had become unavailable. The development of the large-scale aerobic fermentation process for the production of penicillin in deep agitated tanks, which involves the difficult separation of very low concentrations of the antibiotic from the fermentation broth, was carried out under wartime pressure in the early 1940s and is generally recognized as one of the outstanding engineering achievements of the century. The production of chemicals by biological routes remains a core part of *biochemical engineering*, which has always been an essential component of chemical engineering. The discovery of recombinant DNA routes to chemical synthesis has greatly widened the scope of the applications available to the biochemically inclined chemical engineer, and biochemistry and molecular and cell biology have joined physical and organic chemistry, physics, and mathematics as core scientific foundations for chemical engineers.

War is, unfortunately, a recurring theme in identifying the great chemical engineering advances in the twentieth century. The Japanese conquest of the rubber plantations of southeast Asia at the start of World War II necessitated the industrial development of synthetic rubber, and a U.S.-government-sponsored industrial-academic consortium set out in 1942 to produce large amounts of GR-S rubber, a polymer consisting of 75% butadiene and 25% styrene. The chemists and chemical engineers in the consortium improved the production of butadiene, increased the rate of polymerization of the butadiene-styrene molecule, controlled the molecular weight and molecular-weight distribution of the polymer, and developed additives that enabled the synthetic rubber to be processed on conventional natural rubber machinery. By 1945, the United States was producing 920,000 tons of synthetic rubber annually. The synthetic rubber project was the forerunner of the modern synthetic polymer industry, with a range of materials that are ubiquitous in every aspect of modern life, from plastic bags and automobile hoods to high-performance fibers that are stronger on a unit weight basis than steel. Chemical engineers continue to play a central role in the manufacture and processing of polymeric materials.

This short list is far from complete, but it serves our purpose. The chemical engineer of the first half of the twentieth century was generally concerned with the large-scale production of chemicals, usually through classical chemical synthesis but sometimes through biochemical synthesis. The profession began to expand considerably in outlook during the second half of the century.

1.3 The Chemical Engineer Today

Chemical engineers play important roles today in every industry and service profession in which chemistry or biology is a factor, including semiconductors, nanotechnology, food, agriculture, environmental control, pharmaceuticals, energy, personal care products, finance, medicine – and, of course, traditional chemicals and petrochemicals. More than half of the Fourteen Grand Challenges for Engineering in the accompanying block posed by the National Academy of Engineering in 2008 require the active participation and leadership of chemical engineers. Rather than attempt to give a broad picture, we will focus on a small number of applications areas and key individuals. Chemical engineers have traditionally been involved in both the design of *processes* and the design of *products* (although sometimes the product cannot be separated from the process). We include chemical engineers involved with both products and processes, but the entrepreneurial nature of businesses makes it easier to single out individuals who have contributed to products.

The Fourteen Grand Challenges for Engineering

as posed by the U.S. National Academy of Engineering in 2008, prioritized through an online survey.

1. Make solar energy economical
2. Provide energy from fusion
3. Provide access to clean water
4. Reverse-engineer the brain
5. Advance personalized learning
6. Develop carbon sequestration methods
7. Engineer the tools of scientific discovery
8. Restore and improve urban infrastructure
9. Advance health informatics
10. Prevent nuclear terror
11. Engineer better medicines
12. Enhance virtual reality
13. Manage the nitrogen cycle
14. Secure cyberspace

1.3.1 Computer Chips

Andrew Grove

The production of semiconductors is driven by chemical engineers, who have devised many of the processes for the manufacture of computer chips, which are dependent on chemical and rate processes. No one has been more influential in this world-changing technology than Andrew Grove, a chemical engineer who was one of the three founders of the Intel Corporation and its CEO for many years. Grove was selected in 1997 as *Time* Magazine's "Man of the Year." One of the most interesting aspects of Grove's career is that his chemical engineering education at both the BS and PhD levels was a classical one that took place before semiconductor technology could form a part of the chemical engineering curriculum, as it does today in many schools. Hence, it was the fundamentals that underlie the education of a chemical engineer (and, of course, his extraordinary ability) that enabled him to move into a new area of technology and to become an intellectual leader who helped to change the face of civilization.

1.3.2 Controlled Drug Release

Polymer gels that release a drug over time have been investigated since the 1960s. The key issues in timed release are the solubility of the drug in the gel, the uniformity of the rate of release, and, of course, the biocompatibility for any materials placed in the body. One of the leaders in developing this field was chemical engineer Alan Michaels, who was the President of ALZA Research in the 1970s, where he developed a variety of drug delivery devices, including one for transdermal delivery (popularly known as "the patch"). More recently, in 1996, the U.S. Food and Drug Administration (FDA) approved a controlled release therapy for glioblastoma multiforme, the most common form of primary brain cancer, developed by chemical engineer Robert Langer and his colleagues. In this therapy, small polymer wafers containing the chemotherapy agent are placed directly at the tumor site following surgery. The wafers, which are made of a new biocompatible polymer, gradually dissolve, releasing the agent where it is needed and avoiding the problem of getting the drug across the blood-brain barrier. This therapy, which is in clinical use, was the first new major brain cancer treatment approved by the FDA in more than two decades and has been shown to have a positive effect on survival rates. The methodologies used by Michaels, Langer, and their colleagues in this area are the same as those used by chemical engineers working in many other application fields.

Alan Michaels

Robert Langer

1.3.3 Synthetic Biology

Chemical engineers have always been involved in chemical synthesis, but the new field of synthetic biology is something quite different. Synthetic biology employs the new access to the genetic code and synthetic DNA to create novel chemical building blocks by changing the metabolic pathways in cells, which then function as micro-chemical reactors. One of the leading figures in this new field is chemical engineer Jay Keasling, whose accomplishments include constructing a practical and

 inexpensive synthetic biology route to *artemesinin*, which is the medication of choice for combating malaria that is resistant to quinine and its derivatives. Keasling's synthetic process is being implemented on a large scale, and it promises to provide widespread access to a drug that will save millions of lives annually in the poorest parts of the globe. Keasling is now the head of the U.S. Department of Energy's Joint BioEnergy Institute, a partnership of three national laboratories and three research universities, where similar synthetic biology techniques are being brought to bear on the manufacture of new fuel sources that will emit little or no greenhouse gas.

Jay Keasling

1.3.4 Environmental Control

Control of the environment, both through the development of "green" processes and improved methods of dealing with air and water quality, has long been of interest to chemical engineers. Chemical engineer John Seinfeld and his colleagues developed the first mathematical models of air pollution in 1972, and they have remained the leaders in the development of urban and regional models of atmospheric pollution, especially the processes that form ozone and aerosols. The use of Seinfeld's modeling work is incorporated into the U.S. Federal Clean Air Act.

David Boger, a chemical engineer who specializes in the flow of complex liquids (colloidal suspensions, polymers, etc.), attacked the problem of disposing of bauxite residue wastes from the aluminum manufacturing process, which are in the form of a caustic colloidal suspension known as "red mud" that had been traditionally dumped into lagoons occupying hundreds of acres. Boger and his colleagues showed that they could turn the suspension into a material that will flow as a paste by tuning the flow properties (the *rheology*) of the suspension, permitting recovery of most of the water for reuse and reducing the volume of waste by a factor of two. The aluminum industry in Australia alone saves US$7.4M (million) annually through this process, which is now employed in much of the industry worldwide. An environmental disaster in Hungary in 2010, in which the retaining walls of a lagoon containing a dilute caustic red mud suspension collapsed, devastating the surrounding countryside, could probably have been averted or mitigated if Boger's technology had been employed.

John Seinfeld   David Boger

1.3.5 Nanotechnology

Nanotechnology, the exploitation of processes that occur over length scales of the order of 100 nanometers (10^{-7} meters) or less, has been the focus of scientific interest since the early 1990s, largely driven by the discovery of carbon nanotubes and "buckyballs" and the realization that clusters containing a small number of molecules can have very different physical and chemical properties from molar quantities (10^{23} molecules) of the same material. The nanoscale was not new to chemical engineers, who had long been interested in the catalytic properties of materials and in interfacial phenomena between unlike materials, both of which are determined at the nanoscale.

One area in which nanotechnology holds great promise is the development of chemical sensors. As a sensor element is reduced in size to molecular dimensions, it becomes possible to detect even a single analyte molecule. Chemical engineer Michael Strano, for example, has pioneered the use of carbon nanotubes to create nanochannels that only permit the passage of ions with a positive charge, enabling the observation of individual ions dissolved in water at room temperature. Such nanochannels could detect very low levels of impurities such as arsenic in drinking water, since individual ions can be identified by the time that it takes to pass through the nanochannel. Strano has also used carbon nanotubes wrapped in a polymer that is sensitive to glucose concentrations to develop a prototype glucose sensor, in which the nanotubes fluoresce in a quantitative way when exposed to near-infrared light. Such a sensor could by adapted into a tattoo "ink" that could be injected into the skin of suffers of Type 1 diabetes to enable rapid blood glucose level readings without the need to prick the skin and draw blood.

Chemical engineer Matteo Pasquali and his colleagues have found a way to process carbon nanotubes to produce high-strength fibers that are electrically conductive; such fibers could greatly reduce the weight of airplane panels, for example, and could be used as lightweight electrical conductors for data transmission (USB cables) as well as for long-distance power delivery. Pasquali's process is similar to that used for the production of high-strength aramid (e.g., Kevlar™ and Twaron™) fibers, which are used in applications such as protective armor but which are nonconductive. He showed that the carbon nanotubes are soluble in strong acids, where the stiff rodlike molecules self-assemble into an aligned nematic liquid crystalline fluid phase. Nematic liquid crystals flow easily and can be spun into continuous fibers with a high degree of molecular orientation in the axial direction, which imparts the high strength, modulus, and conductivity, then solidified by removing the acid. Pasquali and his team have partnered with a major fiber manufacturer to improve and commercialize the spinning process.

Few commercial applications of nanotechnology have been implemented at the time of writing this text. One of the most prominent is the invention and commercialization of the Nano-Care™ process by chemical engineer David Soane, in which cotton fibers are wet with an aqueous suspension of carbon nanowhiskers that are between 1 and 10 nm in length. Upon heating, the water evaporates and the nanowhiskers bond permanently to the cotton fibers. The resulting fibers are highly stain resistant, causing liquids to bead up instead of spreading. The technology is now in widespread use, as are similar technologies developed by Soane for other applications.

Michael Strano   Matteo Pasquale   David Soane

1.3.6 Polymeric Materials

As we noted in Section 1.2, chemical engineers play a significant role in the synthetic polymer industry, both with regard to the development of new materials and their processing to make manufactured objects. Gore-Tex™ film, which was invented by chemical engineer Robert Gore, is a porous film made from poly(tetrafluoroethylene), or PTFE, commonly known by the trade name Teflon™. Gore-Tex "breathes," in that it passes air and water vapor through the small pores but does not permit the passage of liquid water because of the hydrophobic PTFE surface at the pore mouths. The film is widely used in outdoor wear, but it also has found medical application as synthetic blood vessels. The process requires very rapid stretching of the PTFE film, beyond the rates at which such films normally rupture.

Robert Gore

One example that has been nicely documented in the literature is the development of a new transparent plastic, polycyclohexylethylene, by chemical engineers Frank Bates and Glenn Fredrickson and two chemistry colleagues, for use in optical storage media; the need was for a material that could replace polycarbonate, which absorbs light in the frequency range in which the next generation of storage devices is to operate. Fredrickson is a theoretician who works on polymer theory, whereas Bates is an experimentalist who studies physical properties of block copolymers (polymers made up of two monomers that form segments along the polymer chain that are incompatible with each other). Bates and Fredrickson made use of their understanding of the phase separation properties of incompatible blocks of monomers to utilize the incorporation of penta-blocks (five blocks per chain) to convert a brittle glassy material into a tough thermoplastic suitable for disk manufacture. The description of their collaboration with the chemists in the article cited in the Bibliographical Notes is extremely informative.

Frank Bates   Glenn Fredrickson

1.3.7 Colloid Science

Many technologies are based on the processing and behavior of colloidal suspensions, in which the surface chemistry and particle-to-particle interactions determine

Alice Gast

the properties. Interparticle forces are important when particles with characteristic length scales smaller than about one micrometer come within close proximity, as in the red mud studied by David Boger. Concentrated colloidal suspensions can form glasses or even colloidal crystals. (Opals are colloidal crystals.) Chemical engineers have been at the forefront of the development and exploitation of colloid science in a wide range of applications. One example is work by chemical engineer Alice P. Gast, President of Lehigh University. Electrorheology is a phenomenon in which the viscosity of a suspension of colloidal particles containing permanent dipoles increases by orders of magnitude upon application of an electric field. (Magnetorheology is the comparable phenomenon induced by application of a magnetic field.) The possible application to devices such as clutches and suspensions is obvious. Gast and her coworkers showed theoretically how the interactions between the colloidal forces and the electric field determine the magnitude of the electrorheological response.

1.3.8 Tissue Engineering

Tissue engineering is the popular name of the field devoted to restoring or replacing organ functions, typically by constructing biocompatible scaffolding on which cells can grow and differentiate. Many chemical engineers are active in this field, which is at the intersection of chemical and mechanical engineering, polymer chemistry,

Kristi Anseth

cell biology, and medicine. Kristi S. Anseth, for example, who is a Howard Hughes Medical Institute Investigator as well as a Professor of Chemical Engineering, uses photochemistry (light-initiated chemical reactions) to fabricate polymer scaffolds, thus enabling processing under physiological conditions in the presence of cells, tissues, and proteins. Among the applications that she has pursued is the development of an injectable and biodegradable scaffold to support cartilage cells (*chondrocytes*) as they grow to regenerate diseased or damaged cartilaginous tissue.

1.3.9 Water Desalination

Membrane processes for separation are used in a variety of applications, including hemodialysis (the "artificial kidney") and oxygen enrichment. One of the earliest and

most significant applications was
the development of the reverse
osmosis process for water desalination in 1959 by chemical engineers Sidney Loeb and Srinivasa
Sourirajan. In reverse osmosis,
the dissolved electrolyte migrates
through the membrane away from
a pressurized stream of seawa-
ter or brackish water because
the imposed pressure exceeds the
osmotic pressure. Loeb and Souri-

Sidney Loeb (r) and Srinivasa Sourirajan (l)

rajan showed that the key to making the process work was to synthesize an *asymmetric membrane*, in which a very thin submicron "skin" is supported by a thick porous
layer. (The theoretical foundations for creating asymmetric membranes were developed later.) Reverse osmosis processes currently provide more than 6.5 M m^3/day of
potable water worldwide, and nearly all new desalination process installations use
this technology.

1.3.10 Alternative Energy Sources

Chemical engineers have always been deeply
involved in the development of energy sources,
and with the need to move away from traditional fossil fuel the involvement of the profession has deepened. Solar energy for electricity production is one area in which the chemical engineering role has been notable. Efficient photovoltaic solar modules for electric
power generation are very expensive because of
materials and fabrication costs, and one obvious direction has been to incorporate the con-

Fraser Russell

tinuous production methods used in fabricating films for other applications to
the manufacture of solar cells. T. W. Fraser Russell, who coauthored *Introduction to Chemical Engineering Analysis,* from which this text evolved, led a
research and development team for the continuous production of solar cells and
designed a reactor that deposited the semiconductor continuously on a moving
substrate. Today there are commercial scale operations underway for the continuous manufacture of copper-indium-gallium selenide modules on flexible plastic
substrates.

1.3.11 Quantitative Bioscience

Chemical engineers are playing an increasingly important role in modern biology and biomedicine. For example, Rakesh K. Jain, whose entire education is in chemical engineering, is Professor of Radiation Oncology and Director of the Edwin L. Steele Laboratory for Tumor Biology at Harvard Medical School. Jain and his colleagues have focused on the development of vasculature (the network of blood vessels) and transvascular transport in tumors, with an aim toward developing therapies. His work has been widely recognized in the medical community and has changed the thinking about how to deliver drugs to tumors.

Arup K. Chakraborty is a chemical engineer who uses statistical and quantum mechanics to study molecular conformations. Chakraborty has made major contributions to understanding how zeolites ("molecular sieves") function for separation and catalysis and how polymers interact with surfaces, but he has now turned his attention to fundamental problems in biology. He provided the first quantitative and testable explanation of how the immune synapse (the immune system's recognition process) functions, shed light on the mechanisms underlying the digital response of the orchestrators of adaptive immunity (T cells), described how development shapes the T cell repertoire to mount pathogen-specific responses, and, most recently, illuminated how some humans can control the HIV virus. This work has had a profound impact on the direction of immunological research, most recently in gaining insight into the functioning of the immune system in the presence of the HIV virus.

Rakesh Jain   Arup Chakraborty

1.3.12 Public Service

Chemical engineers are often involved in public service. Lisa P. Jackson, for example, was appointed Administrator of the United States Environmental Protection Agency in 2009, where she directs a staff of 17,000 professionals charged with protecting air and water quality, preventing exposure to toxic contamination, and reducing greenhouse gases, with an annual budget of $10 billion. Samuel W. Bodman, III, who began his professional career as a chemical engineering faculty member, served as the United States Secretary of Energy from 2005 through 2008, heading an agency with an annual budget of over $23 billion and over 100,000 employees.

Lisa P. Jackson   Samuel W. Bodman III

Volunteer work to provide expert advice is often done in the United States through service on panels organized by the National Research Council (NRC), which is the research arm of the National Academies of Science and Engineering. Alice Gast, who was introduced before, chaired an NRC panel charged with determining whether the Federal Bureau of Investigation had employed appropriate scientific techniques when it claimed to have identified the person responsible for mailing *Bacillus anthracis* (anthrax) spores that killed five people in 2001. Chemical engineer Arnold Stancell, who spent most of has career in the petroleum industry, was a member of the NRC panel that investigated the causes of the explosion and fire on the Deepwater Horizon drilling rig in the Gulf of Mexico in 2010, which resulted in eleven deaths and the release of more than 4 million barrels of oil into the Gulf over a three-month period before the well, at a water depth of 1,500 meters (5,000 feet) plus 4,000 meters (13,000 feet) further below the seafloor, was successfully capped. Stancell also served on a committee that advised the U.S. Department of Interior on new regulations to improve the safety of offshore drilling.

Arnold Stancell

Stanley Sandler

Stanley Sandler and other chemical engineers served on three successive NRC panels over a five-year period that evaluated processes for destroying stores of armed weapons loaded with mustard agent and two chemical nerve agents, sarin and VX. The destruction of these weapons is feasible by incineration, which is safe and environmentally benign if properly done, but incineration is sometimes not a politically viable option in populated areas, and the U.S. Congress required the army to consider alternate technologies, which is the task that the NRC was asked to carry out. Numerous technologies were evaluated by the panels on which Sandler served.

1.3.13 Other Professions

Chemical engineers have often made use of their educations to practice other professions. It is no surprise that many chemical engineers choose to study medicine after completing an undergraduate chemical engineering degree, or choose to study law, especially patent law. It is less obvious that many chemical engineers choose to enter the financial sector, which has been a large employer.

Adam Osborne, with BS and PhD degrees in chemical engineering, developed the first commercial portable computer, the Osborne 1, which appeared on the market in 1981. The physicist and Nobel Laureate Eugene Wigner, who is often called the "father of nuclear engineering" because of his World War II work on the uranium separation process, was in fact a chemical engineer by education at all degree levels. The physicist Edward Teller, known as the "father of the hydrogen bomb," studied chemical engineering for his first university degree, as did the mathematician John von Neumann, whose contributions ranged from game theory to the (then) new field of digital computation, and the Nobel Laureate chemists Lars Onsager and Linus Pauling. The former Director of Central Intelligence of the United States, chemist John Deutch, also has a BS degree in chemical engineering; so too does the Dean of the Harvard Business School, Nitin Nohria. Many faculty members in university departments of materials science and engineering, biomedical engineering, environmental engineering, and chemistry studied chemical engineering at the BS level, and in many cases at the PhD level as well. Some chemical engineers have left science completely and had successful careers in the arts or business, including the Academy Award-winning film director Frank Capra and the actor Dolph Lundgren. (This list is not intended to suggest that a chemical engineering education is the key to success in all fields. It is simply to suggest that the tools needed to practice chemical engineering are widely applicable throughout the quantitative disciplines, and that chemical engineering is an expansive profession.)

1.3.14 The Author

As the author of this text, I come with a point of view based on my own experiences as a chemical engineer, and it is useful to comment on these briefly. My formal education is entirely in chemical engineering. I have worked during the course of my career on process optimization and control, fluid mechanics, the analysis of coal gasification reactors for the production of synthetic fuels, the rheology of complex fluids, polymer melt processing (e.g., extrusion and textile fiber manufacture), as well as other areas. I have served as the Editor of the *AIChE Journal,* the flagship journal of the American Institute of Chemical Engineers, and as the Editor of the *Journal of Rheology.* At the time of completing this text I am serving on a National Research Council panel charged with evaluating the methodology of testing body armor for use by the U.S. Army, and I have served on other NRC panels, advisory committees at national laboratories, and so forth. As Director of the Benjamin Levich Institute at the City College of New York I focus on the mechanics and applications of "soft materials;" that is, noncrystalline materials and complex fluids in which the microstructure (colloidal, liquid crystalline, entangled polymer, etc.) plays a large role in determining the properties. I have a joint appointment as Professor of Chemical Engineering and Professor of Physics.

1.4 The Essential Tools

The remainder of this text is devoted to developing the tools used by chemical engineers for the analysis of processes of all types – chemical, physical, or biological. These are the tools used by the practitioners cited in the preceding section, as well as by most members of the profession. Our approach is sometimes called *mathematical modeling,* because we seek to refine the skills required to transform a problem involving physical and chemical phenomena into quantitative form. Mathematical modeling is in some ways an unfortunate name, for the methodology depends on the physical and biological sciences far more than on mathematics, and the mathematical tools required are in fact quite modest; throughout the text we assume only that the reader is familiar with the basic concepts of differential and integral calculus at the level taught in a first course. We are generally dealing with rates in all that we do, so the calculus is the essential *language* that we use for analysis, and it is necessary to become comfortable with that language. (Recall that Newton and Leibniz invented the calculus so that they could attack problems with changing rates.)

The basic approach, which is outlined in the next chapter, is to use the conservation principles of physics – conservation of mass, momentum, and energy – to construct the set of equations that describe the situation of interest. We will concentrate in this text on mass conservation, and, to a lesser extent, on energy conservation, and we will find that we can address a number of realistic problems of considerable inherent interest while developing the necessary methodology.

We cannot, of course, address the scope of problems mentioned in the preceding section, but the student who has mastered the skills that we set out to cover will find that, with further study, all of the areas described previously and more are open.

Bibliographical Notes

Some of the topics described before are addressed in publications that are accessible to the general scientific reader, and it is very important to develop the habit of going to the scientific periodical literature and scientific monographs.

Some of my own thoughts about the profession and its development, which are now more than twenty years old but perhaps still somewhat relevant, are in an essay that was prepared for a symposium noting the 100th anniversary of the first chemical engineering program in the United States. The recorded discussion following the presentation is illuminating. It was here that the definition of chemical engineering that starts this chapter was introduced:

> Denn, M. M., "The Identity of Our Profession," in C. K. Colton, ed., *Perspectives in Chemical Engineering: Research and Education* (*Advances in Chemical Engineering, vol. 16*) Academic Press, New York, 1991.

Two encyclopedias that deal with history, chemistry, and manufacturing operations that are worth browsing, both available in updated electronic editions, are

> *Kirk-Othmer Encyclopedia of Chemical Technology*, 5th Ed., Wiley-Interscience, New York, 2005.
> *Ullman's Encyclopedia of Industrial Chemistry*, 5th Ed., Wiley-VCH, New York, 2005.

A single-volume text that addresses the traditional industries at a level that can be understood with only the background in basic chemistry expected of readers of this book is the most recent edition of the classic *The Chemical Process Industries* by R. N. Shreve,

> Austin, G. T., *Shreve's Chemical Process Industries*, 5th Ed., Mc-Graw-Hill, New York, 1984.

The technical program of the Annual (fall) Meeting of the American Institute of Chemical Engineers contains hundreds of sessions on all aspects of chemical engineering and provides a good overview of the issues of current concern. The program can be found on the Institute's Web site for several months prior to the meeting.

The process for manufacturing computer chips is described in

> Barrett, C. R., "From Sand to Silicon: Manufacturing an Integrated Circuit," *Scientific American Special Issue: The Solid State Century*, January **22**, 1998, pp. 56–61.

Andrew Grove's pioneering text on the subject is

> Grove, A. S., *Physics and Technology of Semiconductor Devices*, Wiley, New York, 1967.

Grove has written several books on business topics. He discusses his life in a memoir:

> Grove, A. S., *Swimming Across: A Memoir*, Warner Books, New York, 2001.

The various physical and chemical steps that the chemical engineer must address in the chip manufacturing process are nicely illustrated in an animated online presentation that is available at the time of writing:

> "How to make a chip," http://www.appliedmaterials.com/HTMAC/animated.html

Web sites should generally be considered to be unreliable sources of information unless those posting the material are well known and there is evidence that the contents have been properly reviewed. Nearly all professional journals use "peer review," in which articles are carefully reviewed by experts to ensure that the results are reliable. Peer review is the reason that scientists and engineers publish their work in professional journals, rather than simply posting it on Web sites. (Review articles, such as those referenced in this section, are sometimes published without peer review, but the authors are carefully selected by the journal editors to ensure accuracy and absence of bias.)

The development of the Weizmann process for acetone production is described in the first sections of

> Jones, D. T., and D. R. Woods, "Acetone-butanol fermentation revisited," *Microbiol Rev.*, **50**, 484–524 (1986).

The penicillin story is the subject of a collection of papers in

> "The history of penicillin production," *Chemical Engineering Progress Symposium Series*, **66**, No. 100 (1970).

For a nice review on controlled drug release, see

> Langer, R., "Drug delivery and targeting," *Nature*, **392** (Supp): 5–10 (1998).

Synthetic biology and Jay Keasling's accomplishments are discussed in the popular press in

> Specter, M., "A life of its own: Where will synthetic biology take us?" *The New Yorker*, September **28**, 2009.

A good overview article directed to a general scientific audience is

> Baker, D., G. Church, J. Collins, D. Endy, J. Jacobson, J. Keasling, P. Modrich, C. Smolke, and R. Weiss, "Engineering life: Building a FAB for biology," *Scientific American*, **294**, 44–51 (June, 2006).

The environmental control topics mentioned in the text are described in

> Seinfeld, J. H., "Air Pollution: A Half Century of Progress," *AIChE Journal*, **50**, 1096–1108 (2004).
> Nguyen, Q. D., and D. V. Boger, "Application of rheology to solving tailings disposal problems," *Int. J. Mineral Processing*, **54**, 217–233 (1998).

Seinfeld has written a basic text on air quality that is designed for an upperclass course:

> Seinfeld, J. H., and S. N. Pandis, *Atmospheric Chemistry and Physics: From Air Pollution to Climate Change*, 2nd Ed., Wiley-Interscience, New York, 2006.

Water quality issues that a chemical engineer might address are included in

> Cech, T. V., *Principles of Water Resources: History, Development, Management, and Policy*, 3rd Ed., Wiley, New York, 2009.

For an introduction to carbon nanotubes, see

> Ebbesen, T. W., "Carbon nanotubes," *Physics Today*, **49**, 26–32 (June, 1996).

Strano's work on nanotechnology is described in

> Lee, C. Y., W. Choi, J.-H. Han, and M. S. Strano, "Coherence resonance in a single-walled carbon nanotube ion channel," *Science*, **329**, 1320–24 (2010).
> Barone, P. W., H. Yoon, R. Ortiz-Garcia, J. Zhang, J.-H. Ahn, J-H. Kim, and M. S. Strano, "Modulation of single-walled carbon nanotube photoluminescence by hydrogel swelling," *ACS Nano*, **3**, 3869–77 (2009).

The processing of carbon nanotubes into fibers is described in

> Behabtu, N., M. J. Green, and M. Pasquali, "Carbon nanotube-based neat fibers," *Nanotoday*, **3**, No. 5–6, 24–34 (2008).

A more detailed scientific treatment of the fiber process is contained in the following article, including a report of improved physical properties:

Davis, V. A., A. N. G. Parra-Vasquez, M. J. Green, P. K. Rai, N. Behabtu, V. Prieto, R. D. Booker, J. Schmidt, E. Kesselman, W. Zhou, H. Fan, W. W. Adams, R. H. Hauge, J. E. Fischer, Y. Cohen, Y. Talmon, R. E. Smalley, and M. Pasquali, "True assemblies of single-walled carbon nanotubes for assembly into macroscopic materials," *Nature Nanotechnology*, **4**, 830–834 (2009).

For an introduction to the use of nanotechnology in textile processing, see

Qian, L., and J. P. Hinestroza, "Application of nanotechnology for high performance textiles," *J. Textile Apparel Tech. Management*, **4**, 1 (2004).

The patent literature is often a good source of information, although patents can be difficult to read because authors often work hard to minimize the amount of information that is revealed about the product or process. Patents can be found through online searches at the Web site of the U.S. Patent Office. The basic patent for David Soane's work on textiles is

Soane, D. W., "Nanoparticle-based permanent treatments for textiles," United States Patent 6607794, 2003.

The literature on nanotechnology is growing at an exponential rate, and new specialized journals have been established. Any general references that we might give at the time of writing are likely to be out of date by the time of publication, and we shall not attempt to do so.

The basic patent for Gore-Tex is

Gore, R. W., "Process for producing porous products," United States Patent 3953566, 1976.

The development of polycyclohexylethylene for storage devices is described in a very readable short article:

Bates, F. S., G. H. Fredrickson, D. Hucul, and S. F. Hahn, "PCHE-based pentablock copolymers: Evolution of a new plastic," *AIChE Journal*, **47**, 762–765 (2004).

There is a nice introduction to block copolymers in

Bates, F. S., and G. H. Fredrickson, "Block copolymers – designer soft materials," *Physics Today*, **52**(2), 32–38 (1999).

There are many good introductions to colloid and surface science, but they presuppose a background in physical chemistry. Three written by professors of chemical engineering are

Adamson, A. W., and A. P. Gast, *Physical Chemistry of Surfaces*, 6th Ed., Wiley-Interscience, 1997.

Evans, D. F., and H. Wennerström, *The Colloidal Domain: Where Physics, Chemistry, Biology, and Technology Meet*, 2nd Ed., Wiley, 1999.

Israelachvili, J. N., *Intermolecular and Surface Forces, with Applications to Colloidal and Biological Systems*, 2nd Ed., Academic Press, 1992.

For an overview of the role of colloid science in electrorheology, see

Gast, A. P., and C. F. Zukoski, "Electrorheological fluids as colloidal suspensions," *Advances in Colloid Science*, **30**, 153 (1989).

Kristi Anseth's work on scaffolding is described in

Cushing, M. C., and K. S. Anseth, "Hydrogel Cell Cultures," *Science*, **316**, 1133–34 (2007).

A recent *Macromolecules Perspective* article on scaffolding is

Shoichet, M. S., "Polymer scaffolds for biomaterials applications," *Macromolecules*, **43**, 581–591 (2010).

For an overview of membrane preparation and the significance of the work of Loeb and Sourirajan, see

Pinnau, I., "Membrane separations: Membrane preparation," *Encyclopedia of Separation Science*, Elsevier, 2000, pp. 1755–1764.

An introduction to sustainable energy written by chemical engineers is

Tester, J. W., E. M. Drake, M. J. Driscoll, M. W. Golay, and W. A. Peters, *Sustainable Energy: Choosing Among Options*, MIT Press, 2005.

Russell's contributions are described in a U.S. patent and references therein:

Wendt, R. G., G. M. Hanket, R. W. Birkmire, T. W. F. Russell, and S. Wiedeman, "Fabrication of thin-film, flexible photovoltaic module," United States Patent 6372538, 2002.

For a readable review of Rakesh Jain's work on drug delivery to tumors, see

Jain, R. K., "Normalization of tumor vasculature: An emerging concept in antiangionic therapy," *Nature*, **307**, 58–62 (2005).

Arup Chakraborty's work on the immune response is described, within a broader context, in

Chakraborty, A. K., and A. Košmrlj, "Statistical mechanical aspects in immunology, *Annual Review of Physical Chemistry*, **61**, 283–303 (2010).

Chakraborty, A. K., and J. Das, "Pairing computation with experimentation: a powerful coupling for understanding T cell signaling," *Nature Reviews Immunology*, **10**, 59–71 (2010).

His initial work on the HIV virus is in

Košmrlj, A., E. Read, Y. Qi, T. M. Allen, M. Altfeld, S. G. Deeks, F. Pereyra, M. Carrington, B. D. Walker, and A. K. Chakraborty, "Effects of thymic selection of the T-cell repertoire on HLA class I-associated control of HIV infection," *Nature*, **465**, 350–354 (2010).

All National Research Council panels, including those mentioned here, issue reports that are peer reviewed prior to release. NRC panel reports are published by the National Academies Press and are available for free downloading at the Council's Web site, http://www.nationalacademies.org.

Some recent overviews that describe work in which I have been involved include the following:

Bonn, D., and M. M. Denn, "Yield stress fluids slowly yield to analysis," *Science*, **324**, 1401–1402 (2009).

Denn, M. M., "Simulation of Polymer Melt Processing," *AIChE Journal*, **55**, 1641–1647 (2009).

There are descriptive chapters describing my earlier work on coal gasification reactors, polymer fiber spinning, and the activated sludge wastewater process in

Denn, M. M., *Process Modeling*, Longman, London and Wiley, New York, 1986.

PROBLEMS

The material in this chapter does not lend itself to typical quantitative problems, but there is a great deal that can be usefully done to amplify on what has been addressed here. Some suggestions follow:

1.1. Select a chemical engineer whose work looks interesting to you, and do a search on his/her publications to get a broader picture. (Your library will have access to several scientific search engines. The Thomson Reuters Web of Science is an excellent place to begin. Its coverage is considerably broader than Google Scholar, but the latter is open access. You should also use the person's home page as a starting point if one exists.)

1.2. Go to your own chemical engineering department's home page and see what kind of scholarly work your faculty members are doing.

1.3. Select a topic of interest to you that involves chemical engineering, do some reading, and write a short review of the outstanding issues. Here are a few suggestions of very broad and socially important topics; they will need to be narrowed considerably for your current purposes: water quality, air quality, global climate change, biofuels, CO_2 sequestration, solar cells for power, energy storage, nuclear waste disposal, targeted drug delivery, nanotechnology, scaffolding for artificial organs.

1.4. Select a chemical that interests you and learn what you can about its production and uses.

1.5. Select a process that interests you and learn what you can about its creation and subsequent development. (Fluid catalytic cracking is a good choice if you don't have another.)

2 Basic Concepts of Analysis

2.1 Introduction

Chemical engineering design, operation, and discovery generally require the analysis of complex physicochemical processes. The quantitative treatment of such systems is frequently called *modeling*, which is a process by which we employ the principles of chemistry, biochemistry, and physics to obtain mathematical equations describing the process. These equations can then be manipulated to predict what will happen under given circumstances. Thus, if it is a chemical reactor that we are modeling, we will know, for example, the effect on the final product of changing the temperature at which the reactor operates. If it is an artificial kidney that we are modeling, we will know the time required for treatment in terms of the flow rate of the dialysis fluid. The analysis process is straightforward and systematic. In this chapter we will examine the approach, see how a model of one simple process unit can be obtained and applied, and get a preview of the things to look for in more complex situations.

2.2 The Analysis Process

The specific goals of analysis are as follows:

1. Describe the physical situation through equations (obtain the model).
2. Use the model equations to predict behavior.
3. Compare the prediction with the actual behavior of the real system.
4. Evaluate the limitations of the model, and revise if necessary.
5. Use the model for prediction and design.

The logical sequence of the analysis process is shown in Figure 2.1. This is a manifestation of what is often called the *scientific method.*

The physical situations that are of interest to chemical engineers include the behavior of objects as diverse as equipment, such as chemical or biochemical reactors, heat exchangers, and distillation columns; rivers and estuaries; biological cells; and organs or entire organisms. We might need a mathematical description of the properties of a material – perhaps a porous membrane in terms of its composition and

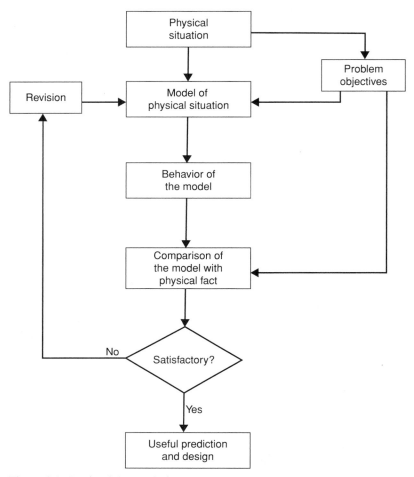

Figure 2.1. Logic of the analysis process.

preparation. Whether we are trying to describe the behavior of a piece of equipment, a part of the human circulatory system, or any other physicochemical phenomenon, the development of a mathematical model proceeds in the same manner.

The basic sources of any mathematical description are the conservation principles for mass, energy, and momentum. Taken together with other fundamental principles of physics, such as gravitational attraction, it seems possible in principle to obtain a mathematical description of any physicochemical phenomenon. That this is an unreasonable expectation in fact is obvious at once, for, although nineteenth-century scientists thought that such an outcome was just beyond the horizon, physics is still a very active science. (Simply recall from the basic physics course the complexity of describing the state of a single gas in terms of the individual behavior of 10^{23} interacting molecules.) Thus, we may expect that there will be many situations of engineering interest that are too complex for the laws of physics to be applied in their most fundamental form. We therefore need a secondary source from which to draw to develop mathematical models. This nonfundamental source, so essential to engineering analysis, produces what we call *constitutive relationships.* Constitutive relationships are generally developed from careful and clever experimentation for specific situations of interest. (The term originated in the field of the mechanics of materials, where the word *constitutive* indicates that the relation is not general, but is specific to a particular *material constitution.*) Development of a systematic approach to mathematical description using the conservation laws and constitutive relationships is a major concern of this text, and much of what follows is devoted to meeting this goal.

Most mathematical descriptions will represent an essential compromise between the complexity required for description of a physical situation that is true in every detail and the simplicity required so that the model may be compared with experiment and then used for design and operation. The degree of compromise depends on the specific problem objectives and often determines the effort that we devote to obtaining a model.

Given the mathematical description, it is necessary to verify that it is correct before using it for any engineering purpose. This step is often called *model validation,* and it has occupied the attention of scientists and philosophers of science for decades. Model validation requires solving the equations to predict the behavior of the mathematical model under conditions where a direct comparison can be made with the behavior of the real physical situation. The challenge in model validation is to ensure that the comparison is one that truly tests the model. (We will see a very elementary example of this challenge later in this chapter.) It is during model validation that the engineer makes value judgments about the usefulness and reliability of a model for subsequent design and prediction. If, for a given set of objectives, the comparison between model and physical reality is adequate, then we may proceed to use the model; if not, we must consider why the comparison is inadequate, make appropriate modifications, and compare again.

2.3 Source of the Model Equations

A procedure for constructing a mathematical model for an extremely simple physical situation is shown in Figure 2.2. We presume for definiteness that we are seeking to describe the behavior of a piece of process equipment consisting of a tank that has liquid streams flowing in and out. The first step is the selection of what we will call *fundamental dependent variables.* The fundamental dependent variables are the collection of quantities whose values at any time contain all of the information necessary to describe the process behavior. There are only three such fundamental variables in most problems of interest to us: mass, energy, and momentum.

In many instances the fundamental dependent variables cannot be conveniently measured. We do not have an energy meter, for example; rather, we deduce the energy of a system by knowing the temperature, pressure, composition, and so forth. Similarly, it is likely that we will deduce mass from measurements of density, volume, and so on, whereas momentum will be deduced from measurements of velocity and force. These *characterizing dependent variables* are variables that can be conveniently

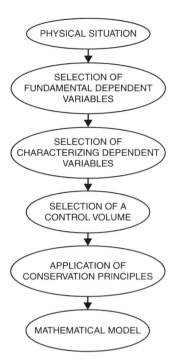

Figure 2.2. Model development for simple situations.

measured and, properly grouped, determine the values of the fundamental variables. Generally, more than one characterizing variable (density, temperature, pressure, flow rate, composition, etc.) is needed to specify each fundamental variable. The values of all the characterizing variables at any time and at any point in space define the *state* of the system, and characterizing variables are called *state variables* in the field of process dynamics and control. (*State variable* has a more restricted meaning in thermodynamics.)

There are four independent variables of concern to us in engineering problems: time (t) and the three coordinates that establish position in space; in rectangular Cartesian coordinates the spatial variables are usually denoted x, y, and z. In any given situation we may be concerned with the system behavior with respect to changes in time and space; the focus in this introductory text will be on time dependence, with the occasional look at variation in one spatial dimension. Our task is now to establish a systematic procedure for selecting the pertinent dependent and independent variables and utilizing the conservation laws.

2.4 Conservation Equations

The first quantitative step in model development is the application of the conservation principles. This step, which we shall discuss in some detail, produces the basic model equations for the physical situation. The conservation laws are bookkeeping statements (*balance equations*) that account for mass, energy, or momentum.

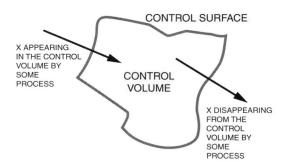

Figure 2.3. Control volume.

Consider any region of space enclosed for the purpose of "accounting" by a (fictitious) surface, which we call the *control surface*; the volume within the control surface is called the *control volume*. A control volume is shown schematically in Figure 2.3. We designate the quantity to be conserved as X, which may be mass, momentum, or energy. The conservation law may be stated as follows:

The total amount of X contained within the control volume at some time t_2 is equal to the total amount of X contained within the control volume at an earlier time t_1, plus the total amount of X that has appeared in the control volume by all processes in the time interval from t_1 to t_2, less the total amount of X that has disappeared from the control volume by all processes in the time interval from t_1 to t_2.

As we shall see shortly, we almost always chose t_1 and t_2 to be close together, because we wish to utilize differential calculus by taking the limit as $t_2 \to t_1$. Thus, denoting t_1 by t and t_2 by $t + \Delta t$, we write the conservation equation as

$$X|_{t+\Delta t} = X|_t + \text{amount of } X \text{ entering during } (t, t + \Delta t)$$
$$- \text{amount of } X \text{ leaving during } (t, t + \Delta t). \tag{2.1}$$

The symbol "$|_t$" denotes "evaluated at t." t is a given time for purposes of writing the equation, but it can be *any* time.

The amount of X entering the control volume during the interval $(t, t + \Delta t)$ equals the *rate* at which X enters (quantity/time) multiplied by the time interval:

amount of X entering during $(t, t + \Delta t) = [\text{rate at which } X \text{ enters}] \, \Delta t$

A similar equation applies for the amount leaving. The rates may be different at different times, but (as is always the case in applying the calculus) we presume that the rate is constant over the vanishingly small interval Δt. The word statement is then

$$X|_{t+\Delta t} = X|_t + [\text{rate at which } X \text{ enters} - \text{rate at which } X \text{ leaves}] \, \Delta t, \tag{2.2a}$$

or

$$\frac{X|_{t+\Delta t} - X|_t}{\Delta t} = \text{rate at which } X \text{ enters} - \text{rate at which } X \text{ leaves}. \tag{2.2b}$$

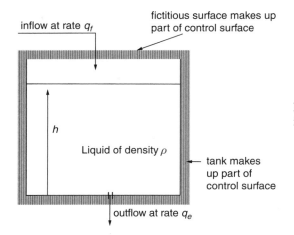

Figure 2.4. A well-mixed tank with inflow and outflow.

We now take the limit as $\Delta t \to 0$. The *difference quotient* on the left-hand side of the equation becomes the derivative, and we obtain

$$\frac{dX}{dt} = \text{rate at which } X \text{ enters} - \text{rate at which } X \text{ leaves.} \qquad (2.3)$$

The derivative has the physical meaning of rate of change, and we thus find that a completely equivalent statement of the conservation law is as follows:

> *The rate of change of the total amount of X contained within the control volume at any time t is equal to the rate at which X enters the control volume at time t by all processes, less the rate at which X leaves the control volume at time t by all processes.*

Our task is now as outlined in Figure 2.2. Suppose that we have a physical problem of interest. (We will consider several specific processes later in this chapter, and others of increasing complexity throughout the text.) We must then decide which conservation equations are relevant, for it is the conservation equations that determine the fundamental dependent variables. Next, we must identify the variables that characterize the fundamental variables. Finally, we must select the control volume in order to apply the conservation principles.

2.5 An Application of Mass Conservation

Many problems of interest involve processes that take place in well-mixed systems with inflow and outflow. The system might be a tank used as a mixer, or a reactor for a chemical or petrochemical process, or it might be a small bioreactor for the production of high-value proteins from microorganisms. Many metabolic processes in organs can be described by treating the organ as one or more well-mixed control volumes with inflow and outflow.

Consider the cylindrical tank shown schematically in Figure 2.4. The tank contains a liquid of density ρ. The same liquid flows in and out of the tank. The cross-sectional area is A, and the height of liquid in the tank at any time t is $h(t)$. The inflow is at volumetric flow rate q_f, measured, for example, in m³/sec, or perhaps

ft^3/min. The outflow is at a rate q_e. (The subscripts f and e denote *feed* and *effluent*, respectively.) The flow rates q_f, q_e, or both, might be changing with time. We assume that the temperature throughout the system is always the same, so we needn't worry about the effect of changing temperature on the density. We wish to understand how the liquid level in the tank changes, and how we might control the level.

The fundamental dependent variable is clearly mass. We could monitor the mass directly by putting the tank on a spring balance, although that would usually be impractical. It is likely that we will choose to characterize the mass by measuring the volume and density, and the mass flow rates by measuring or selecting volumetric flow rates and density. Thus, the variables describing mass conservation are ρ and A (which are constants), h, q_f, and q_e.

It is not always a simple matter to identify the control volume, but in this case the control volume is clearly the tank. The control surface, as shown in Figure 2.4, consists of the tank walls and, if the tank is open to the atmosphere at the top, a fictitious surface separating the contents of the tank from the surroundings.

The total amount of mass in the control volume (the tank) is equal to the density multiplied by the volume, ρAh.* The rates of inflow and outflow are, respectively, ρq_f and ρq_e. The principle of conservation of mass in rate form is then

$$\frac{d}{dt} \rho Ah = \rho q_f - \rho q_e. \qquad (2.4)$$

That is, the rate of change of mass in the control volume equals the rate at which mass enters less the rate at which mass leaves. The density is not changing with time, so it can be taken outside the derivative, and the area A is, of course, a constant. The density cancels from both sides of the equation, and we finally obtain

$$\frac{dh}{dt} = \frac{q_f}{A} - \frac{q_e}{A}. \qquad (2.5)$$

The rate of change of the height with time (dh/dt) is zero if $q_f = q_e$, and the liquid level remains constant. We say a system is at *steady state* when all time rates-of-change (i.e., all time derivatives) are zero. Many systems are designed to operate at steady state.

If there is an inflow, but no outflow ($q_e = 0$), we call the system *semibatch*. Many processes for the production of pharmaceuticals and fine chemicals are semibatch. q_f may be different at different times, perhaps changing continuously. With $q_e = 0$ we can formally write Equation 2.3 in *separated* form, in which everything that depends on h is on the left-hand side and everything that depends on t is on the right-hand side, as

$$dh = \frac{q_f}{A} dt.$$

---

* This neglects the negligible mass of any air that sits above the liquid in the tank. We could avoid this minor complication by taking the upper surface of the liquid as part of the control surface. Nothing in what follows would change.

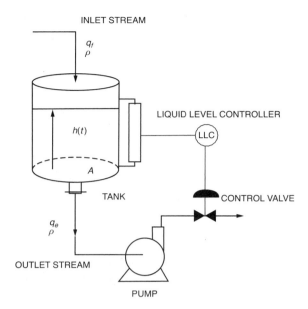

Figure 2.5. Process tank with pump and simple control loop.

Integrating from $t = 0$ to any later time t we obtain

$$h(t) = h_0 + \int_0^t \frac{q_f(\tau)}{A} d\tau,$$

where τ is a "dummy variable" of integration, representing all times between $t = 0$ and the present time. h_0 is the height at $t = 0$; since there is one integration, there will be one constant of integration, hence it is necessary to know the state of the system at one time. The integration can be performed for any function $q_f(t)$; if q_f is a constant, we obtain

$$h(t) = h_0 + \frac{q_f}{A} t.$$

2.6 A Design Problem

2.6.1 Problem Formulation

Few engineering problems result in models as elementary as Equation 2.5, but even this model has one very instructive application. Consider the system shown in Figure 2.5. Suppose that q_f is changing in time, but we wish to maintain the tank level constant at a value h_0 despite these changes in q_f. Suppose also that we can measure the tank level continuously, but that we cannot measure q_f reliably in a continuous manner. We will therefore monitor the liquid level continuously and adjust a valve on the exit line, so we change q_e continuously to compensate for fluctuations in q_f.

Suppose the system is designed to operate at steady state ($dh/dt = 0$) with $q_f = q_e = q^*$, where q^* is a constant, and the desired level is h^*. Now suppose that at time $t = t_1$ there is a step change in q_f, as shown in Figure 2.6a, to a value $q_f = q^* + Q^*$;

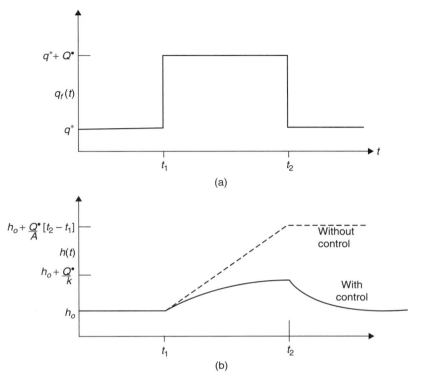

Figure 2.6. Change in feed flow rate (a) and corresponding change in tank volume (b).

q_f remains at this value until time t_2, after which the value again drops to $q_f = q^*$. If q_e remains unchanged, then Equation 2.5 becomes

$$\frac{dh}{dt} = 0, \quad < t_1$$

$$\frac{dh}{dt} = \frac{Q^*}{A}, \quad t_1 \le t \le t_2$$

$$\frac{dh}{dt} = 0, \quad t_2 < t.$$

With $h = h^*$ for $t < t_1$ we can then integrate this equation to obtain

$$h = h^*, \quad t < t_1$$

$$h = h^* + \frac{Q^*}{A}[t - t_1], \quad t_1 \le t \le t_2$$

$$h = h^* + \frac{Q^*}{A}[t_2 - t_1], \quad t_2 < t.$$

Note that the integration is straightforward; if dh/dt is a constant, $h(t)$ must be a straight line with slope Q^*/A. As shown in Figure 2.6, although the flow rate returns to the design value $q_f = q^*$, the level does not return to the design value $h = h^*$.

2.6.2 Feedback Control

We can design a control system to maintain the level within the desired limits by implementing a concept known as *feedback control*. Level controls were probably the earliest form of feedback control systems, and their use has been documented as early as the third century BCE. Feedback control was largely developed during World War II, but there was a notable early industrial application in 1788 by James Watt, who used existing ideas to design a regulator to maintain a constant pressure in steam engines; Watt's design was subsequently analyzed in 1868 in a classic paper by James Clerk Maxwell, the Scottish physicist best known for the development of electromagnetism. Watt's design is similar in concept to the regulator in a kitchen pressure cooker, in which a small weight is placed atop a hole to allow some steam to escape; if the internal pressure is too high, the weight rises to allow more steam to escape and the pressure to fall, whereas if the pressure is too low, the weight lowers to slow the escape of steam and allow pressure to build up. The greater the deviation from the desired pressure, the more the control action that will be taken. In a similar way, we monitor the tank level; we decrease the exit flow rate q_e if the level falls below a desired value, whereas if the level rises above the desired value we increase q_e. The more the deviation from the desired level, the more the control action we require. We will assume that the control action should be *proportional* to the deviation. Thus, we get

$$q_e = q^* + K[h - h^*], \tag{2.6}$$

where K is a constant to be determined.

Now, suppose that the variations in the inlet flow rate can be represented by a function of time $Q(t)$; i.e.,

$$q_f = q^* + Q(t). \tag{2.7}$$

The mass balance, Equation 2.5, then becomes

$$\frac{dh}{dt} = \frac{1}{A} \left\{ \underbrace{q^* + Q(t)}_{q_f} - \underbrace{q^* - K[h - h^*]}_{q_e} \right\}$$

or, equivalently,

$$\frac{dh}{dt} + \frac{K}{A}[h - h^*] = \frac{Q(t)}{A}. \tag{2.8a}$$

Finally, since we are interested in the difference $h - h^*$, and since h^* is a constant, $d[h - h^*]/dt = dh/dt - dh^*/dt = dh/dt$, and we can write

$$\frac{d}{dt}[h - h^*] + \frac{K}{A}[h - h^*] = \frac{Q(t)}{A}. \tag{2.8b}$$

We seek a solution to this equation starting at $t = 0$, where the system is presumed to be at the desired level $h = h^*$.

Equation (2.8) can be integrated by means of an *integrating factor*, which is a standard procedure for equations of this type and is covered in detail in calculus texts. We note that the left-hand side of Equation (2.8) will become the derivative of a single term if we multiply by $e^{Kt/A}$. Hence, we multiply both sides of the equation by $e^{Kt/A}$ to obtain

$$\frac{d}{dt}\left\{e^{Kt/A}[h - h^*]\right\} = \frac{e^{Kt/A}}{A} Q(t).$$

Integrating both sides from $t = 0$ to any time t, and using the fact that $h - h^* = 0$ at $t = 0$, gives

$$h(t) - h^* = \frac{e^{-Kt/A}}{A} \int_0^t e^{K\tau/A} Q(\tau)d\tau, \tag{2.9}$$

where τ is the "dummy variable" denoting all times between the two limits of the integral.

Now let us consider again the step changes in flow rate shown in Figure 2.5a, with $Q(t)$ equal to the nonzero constant value Q^* for $t_1 \leq t \leq t_2$ and zero elsewhere. We leave it as an exercise in integration to show that

$$h - h^* = \begin{cases} 0, & 0 \leq t < t_1 \\ \frac{Q^*}{K}[1 - e^{-K[t-t_1]/A}], & t_1 \leq t < t_2 \\ \frac{Q^*}{K}[e^{-K[t-t_2]/A} - e^{-K[t-t_1]/A}], & t_2 \leq t \end{cases}.$$

The response is shown in Figure 2.6b, where it is compared with the behavior of the system without control $(K = 0)$. The difference is striking, and it is clear that by proper choice of K the fluctuations in tank level can be kept within desired limits.

2.6.3 Controller Design

Thus far we have been engaged in an exercise that allows us to calculate the level response for any given change in the inlet flow rate, with and without feedback control. What we really need is a way of *designing* the control system; that is, of specifying the *feedback gain K* of the controller so that the system always stays within desired tolerances, regardless of the specific nature of the disturbance. We can accomplish this with the model equation. We suppose that fluctuations in the inlet feed will never exceed a fraction f of the total design flow rate, q^*; that is,

$$\text{maximum of } |Q(t)| \leq fq^*.$$

We wish to choose K such that fluctuations in the liquid level never exceed a fraction φ of the design level h^*; that is,

$$\text{maximum of } |h(t) - h^*| \leq \varphi h^*.$$

From Equation (2.9) we can then write the constraint on $|h(t) - h^*|$ as

$$\text{maximum of } \left| \frac{e^{-Kt/A}}{A} \int_0^t e^{K\tau/A} Q(\tau)d\tau \right| \leq \varphi h^*.$$

The left side of this inequality will take on its largest value when $Q(t)$ is as large as possible for all time, so we replace $Q(t)$ in the integral with fq^* and write (noting that we can drop the absolute value signs, since all terms are positive)

$$\frac{e^{-Kt/A}}{A} \int_0^t e^{K\tau/A} fq^* d\tau \leq \varphi h^*.$$

Since fq^* is a constant, the integral is only an exponential, and we carry out the integration to obtain

$$\frac{fq^*}{K} \left[1 - e^{-Kt/A}\right] \leq \varphi h^*.$$

The left side takes on its largest value for $t \to \infty$, so we have

$$\frac{fq^*}{K} \leq \varphi h^*,$$

or, solving for K,

$$K \geq \frac{fq^*}{\varphi h^*}.$$

The smallest value satisfying this inequality will require the least activity on the part of the control system while still satisfying the design specifications. Thus, the solution to our design problem is

$$K = \frac{fq^*}{\varphi h^*}; \tag{2.10a}$$

that is, the minimum feedback gain required to meet the design specifications is the ratio of the maximum expected flow rate perturbation to the maximum acceptable perturbation in height. The exit flow rate is thus required to depend on the liquid level by the equation

$$q_e = q^* \left[1 + \frac{f}{\varphi} \left(\frac{h - h^*}{h^*}\right)\right]. \tag{2.10b}$$

For example, if feed-rate fluctuations can go as high as 10 percent of the design feed flow rate, and it is required that the liquid level stay constant to within 0.15 percent, then $f = 0.10$, $\varphi = 0.0015$, and

$$q_e = q^* \left[1 + 66.7 \left(\frac{h - h^*}{h^*}\right)\right].$$

That is, for each percentage change in measured liquid level, the exit flow rate is increased or decreased by 66.7 percent to ensure that the design tolerance is never exceeded. Note that this is a conservative design, and that generally the fluctuations in h will be less than 0.15 percent.

Figure 2.7 shows the results of a *simulation* (i.e., a solution of the model for a specific case) for $A = 10 \text{ cm}^2$, $h^* = 100 \text{ cm}$, and $q^* = 25 \text{ cm}^3/\text{s}$, with $f = 0.10$ and $\varphi = 0.0015$ ($K = 16.67 \text{ cm}^2/\text{s}$). The disturbance is plotted as $Q(t)/q^*$ in Figure 2.7a, whereas the response is plotted as $[h(t) - h^*]/h^*$ in Figure 2.7b with and without control. It is clear that the controller is effective in maintaining the level within the

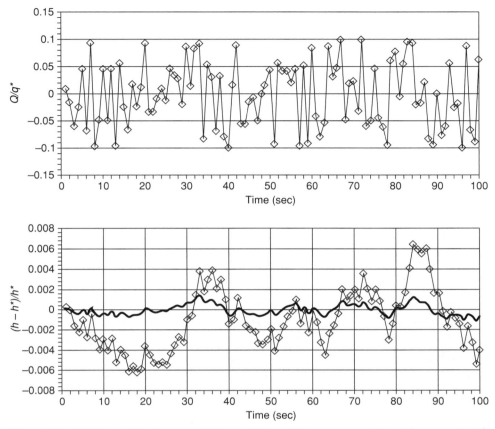

Figure 2.7. (a) Inflow disturbance and (b) simulation of the response without (open symbols) and with (heavy solid line) feedback control. $K = 16.67$ cm^2/s.

specification of a maximum variation in height of 0.15 percent of the design value, despite inflow fluctuations that are sometimes large.

2.6.4 Further Comments

The proportional controller designed here is the type that is routinely used in level control applications. A few questions arise about this particular selection of a design methodology, and it is worth a brief digression here, although the issues will be addressed in a broader context subsequently in the curriculum. First, we may ask why we did not choose the alternative strategy of measuring the changes in q_f and changing q_e so as to keep $dh/dt = 0$ at all times. This strategy is known as *feedforward control,* wherein we attempt to measure the disturbance and compensate for it directly, rather than controlling by measuring the deviation from the desired state. Feedforward control can be an effective strategy, but it has one obvious disadvantage: We never deal directly with the quantity that we wish to control (the level), so any small error in measurement or compensation will result in a change in the level that

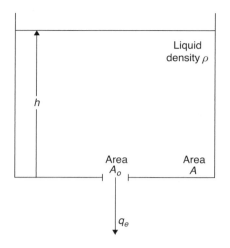

Figure 2.8. Tank draining through an orifice in the bottom.

will never be corrected. In that sense, feedback is a more conservative approach. In reality, a combination of feedforward and feedback is often employed.

Examination of Equation 2.9 shows another concern that might need to be addressed in practice. If the inflow simply changes by a constant amount and never changes back to zero, the height will never return to the value h^*, but will rather go to a steady-state value $h = h^* + Q^*/K$. Q^*/K is known as *steady-state offset*. Steady-state offset is not a problem if the disturbances are continuous and exhibit positive and negative deviations from the design basis. There are straightforward ways to address steady-state offset, but that is beyond the scope of this introductory treatment.

2.7 Are Conservation Equations Sufficient?

2.7.1 The Draining Tank

The example used thus far illustrates the logic in Figure 2.2, but it is deceptive in its simplicity. It is, in fact, unusual that the conservation equations alone lead so directly to a model that can be employed for analysis and design. We illustrate this important point here with a slight variant of the tank problem, as shown in Figure 2.8. There is no inflow ($q_f = 0$); the tank empties by gravity-driven flow through a small hole, or *orifice*, in the base. The area of the orifice is A_0. As before, there is no heating or cooling, and the temperature of the liquid remains constant. Our objectives might be the answer to some or all of the following questions:

How long will it take the tank to drain?

How does the height of liquid vary with time?

How does the flow rate through the orifice vary with the depth of liquid?

This tank draining is a physical situation with which most people have had direct experience; if not, it simply requires punching a hole in a large can, filling it with water, and then observing the behavior as the liquid flows out. Observation shows us that the level in the tank decreases with time, the flow rate of liquid through the orifice varies with the height of the liquid and with the size of the exit orifice, and the tank empties completely in a finite time. Little more can be said with verbal statements, and we turn again to the conservation equation. The situation is identical to that leading to Equation 2.5, so with $q_f = 0$ we obtain

$$\frac{dh}{dt} = -\frac{q_e}{A}. \tag{2.11}$$

In contrast to the preceding example, Equation 2.11 is a single equation involving two quantities that we do not know: the liquid height, h, and the effluent rate, q_e. (Henceforth, since there is only one flow rate and there is no possibility of confusion, we will denote the effluent flow rate simply by q.) Since we have two unknowns and only one equation we must seek a second relation. This relation can, in fact, be established with some approximations from the principle of conservation of momentum or from the principle of conservation of energy, either of which can be used to derive a fundamental relation in fluid mechanics known as the *Bernoulli equation*; indeed, you may have already been introduced to the Bernoulli equation in a physics course. It is often the case, however, that we are unwilling or unable to apply further conservation equations at a convenient level of complexity, and we shall presume somewhat artificially that such is the situation here. Our additional relationship between q and h, the *constitutive relationship*, must then be obtained by intuition and/or experiment. We anticipate that a relationship obtained in this way will be rather less general than one based on fundamental conservation principles, and we must use great care in applying the results to situations that differ very much from the conditions of any experiments that we have performed.

Now, we know that the flow occurs through the orifice because the pressure in the liquid at the base of the tank is greater than the pressure of the atmosphere, thus forcing the liquid out, and that the greater the pressure difference the greater the flow. We can express the general relationship as $q = q(\Delta p)$, by which we mean that q, the flow rate, is a function of Δp, the pressure change across the orifice. If the top of the tank is open, then the pressure there, too, is atmospheric. The pressure in the liquid at the bottom of the tank exceeds the pressure of the atmosphere by the weight per unit area of the liquid column, which is proportional to the height of liquid. The pressure difference, or *driving force* for flow, is therefore proportional to h. It is the functional relationship of q to h, denoted as $q(h)$, that we seek as our second relation to supplement Equation 2.11. The approach that we shall take is to postulate the form of the dependence of q on h (our constitutive relation), solve the model Equation 2.11 for h, and then check the prediction of the model with the experimental data. If the model and data do not agree, we will use the way in which they disagree as an aid in postulating a new dependence. (This is the *Revision* step in Figure 2.1.)

Table 2.1 shows some data of liquid (water) height versus time for three experimental runs in a draining tank. (Height was taken as the independent variable in the experiment, since it is easier to read the time for a given height than vice versa. The

Table 2.1. *Liquid height versus time for the tank emptying experiment. Tank diameter = 27.3 cm (10.75 in.), tank height = 30.5 cm (12 in.), orifice diameter = 1.55 cm (0.61 in.)*

Height of liquid (centimeters)	Time (seconds)	Height of Liquid (centimeters)	Time (seconds)
30.5	0	15.2	36.4
0		35.8	
0		36.4	
27.9	5.8	12.7	43.8
6.1		42.8	
5.9		43.8	
25.4	10.9	10.2	51.0
11.5		50.5	
11.6		51.6	
22.9	16.6	7.6	60.2
17.8		59.2	
17.2		60.6	
20.3	23.0	5.1	71.0
23.5		69.8	
23.0		71.4	
17.8	30.	0	85.0
29.2		84.0	
29.8		85.2	

data were originally recorded in even inches, and the centimeter values are rounded to one decimal place.) The data are plotted in Figure 2.9. In most cases the three data points cannot be distinguished on this scale, and only a single point is shown. There are also two lines in Figure 2.9; the lower line is a straight line drawn through the first three data points and extrapolated, while we will discuss the upper line a bit later. The data clearly indicate that the slope (i.e., the rate of change of height), which is proportional to the flow rate, decreases in magnitude with decreasing height. This is consistent with our understanding of the physical process: We know the liquid will flow out more slowly for a small height than a large one, and there can be no flow if there is no liquid height at all. This last observation seems trivial but in fact has a

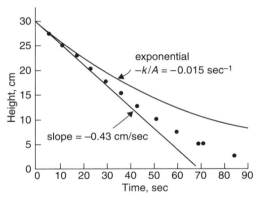

Figure 2.9. Liquid height as a function of time for a draining tank.

profound implication, for it tells us that whatever the relation between q and h may be, q must vary as a positive power of h for sufficiently small h.

2.7.2 Assuming Linearity

It is instructive at this point to follow what will turn out to be a nonproductive path. Early physical scientists, including Leonardo de Vinci and Newton, were often confronted with the type of problem we face here, that of determining a constitutive relation (although Leonardo and Newton did not use these twentieth-century words). Their choice was frequently to assume linearity, which would lead to the form $q = kh$, where k is a constant. Equation 2.11 then simply becomes

$$\frac{dh}{dt} = -\frac{k}{A}h. \tag{2.12}$$

The only function proportional to itself is an exponential, so we know the solution to this first-order, homogeneous ordinary differential equation, but it is useful to work through the details of the solution. The equation is *separable*, and we can follow the shortcut "separation of variables" method to write Equation 2.12 symbolically as

$$\frac{dh}{h} = -\frac{k}{A}dt$$

or, integrating time between $t = 0$ and a later time t, and the height from its value h_0 at time $t = 0$ to its value at t, we have

$$\ln\frac{h(t)}{h_0} = -\frac{kt}{A}. \tag{2.13}$$

Taking the exponential of both sides yields

$$h(t) = h_0 e^{-kt/A}. \tag{2.14}$$

It is always best to compare a model to data on a plot where the data are expected to be linear, since that minimizes the effort and permits us to employ the method of least squares (Appendix 2E) if we wish. In this case, based on Equation 2.13, we plot the natural logarithm of $h(t)/h_0$ versus t, as shown in Figure 2.10. The data are linear only over the first three or four points, so the assumed functional form is clearly incorrect, but that is not surprising; Equation 2.14 predicts that an infinite time is required for the tank to drain, while we know that the tank empties in a finite time. Equation 2.14 is shown as the upper line on Figure 2.9, where the value of k/A used is that obtained by fitting the short-time data in Figure 2.10.

2.7.3 Power Dependence

The two lines in Figure 2.9 bound the actual behavior. The exponential response shows that a linear dependence of q_e on h is too strong. The straight line, corresponding to no dependence, or $q_e = $ constant, is too weak. It often happens that physical phenomena are described by power relations (sometimes called *power laws*), either

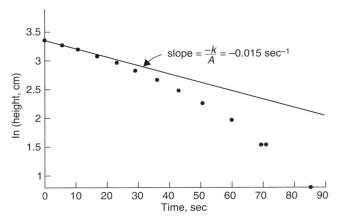

Figure 2.10. Natural logarithm of liquid height versus time. The straight line is Equation 2.10.

for sound theoretical reasons or simply because the power relation fits data well over one or more decades of the independent variable. In the case at hand, a power relation would be of the form

$$q_e = kh^n \tag{2.15}$$

with $0 < n < 1$. Equation 2.11 then becomes

$$\frac{dh}{dt} = -\frac{kh^n}{A} \tag{2.16}$$

or, in separated form,

$$\frac{dh}{h^n} = -\frac{k}{A}dt.$$

Integrating the right side from time zero to the present time, and the left side with respect to h from h_0 at time zero to h at the present, we obtain

$$\frac{h^{1-n}}{1-n} - \frac{h_0^{1-n}}{1-n} = -\frac{kt}{A} \tag{2.17a}$$

or, solving the algebraic equation for h,

$$h(t) = h_0 \left[1 - \frac{k[1-n]}{Ah_0^{1-n}}t \right]^{1/[1-n]}. \tag{2.17b}$$

The system empties in a finite time only for $n < 1$. For $n = 0$ we simply have $q_e = $ constant and we recover the straight line in Figure 2.9. It is readily shown that Equation 2.17b reduces to Equation 2.14 in the limit $n \rightarrow 1$, so, in fact, the power relation includes the two simpler limiting cases.

Now this slight generalization has, in fact, greatly increased the complexity of the analysis, for there are now two parameters, k and n, that must be determined from the experiment. A rational approach might be to choose a value for n, plot h^{1-n} versus t as motivated by Equation 2.17a, and check for linearity, choosing a new value of n according to how the data deviate from linearity. Ultimately we will arrive at the

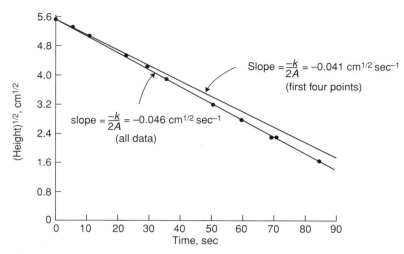

Figure 2.11. Square root of liquid height versus time. The straight lines are Equation 2.14 for $n = 1/2$.

"best" value of n to represent the data and find the corresponding value of k. There is good reason to believe, however, that $n = 1/2$ is the proper value. In Appendix 2A we show by estimating the derivative of h from the data that n must lie between 0.37 and 0.66, with the most likely value close to 0.5. In Appendix 2D we obtain the result $n = 1/2$ using a technique known as *dimensional analysis*, which follows from the requirement that all terms in an equation must have the same dimensions. The topic is also briefly addressed in the next section. With $n = 1/2$, Equation 2.17b becomes

$$h^{1/2} = h_0^{1/2} - \frac{kt}{2A}, \tag{2.18a}$$

$$h = h_0 \left[1 - \frac{kt}{2Ah_0^{1/2}} \right]^2. \tag{2.18b}$$

Figure 2.11 shows a plot of $h^{1/2}$ versus t, and, although there is some scatter to the data, the best line[*] through all points has a slope $-k/2A = -0.046$ cm$^{1/2}$ s^{-1}. We conclude, then, that the flow rate q_e in this experiment is well represented by the relation

$$q_e = 8.3 h^{1/2},$$

where we have used the value of A from Table 2.1. (We can also fit a line to the first four data points on Figure 2.11. If we do this we find that the slope of such a line is $-k/2A = -0.041$, which is slightly different from the value obtained using all data points. This is due to scatter in the data.)

The analysis process has now been illustrated to the point where the particular physical situation has been adequately modeled. We still need to consider how

---

[*] For a discussion of obtaining the best fit to a set of data, see Appendix 2E.

reliable the mathematical description may be when used to predict behavior in similar systems for purposes of design. We know nothing, for example, about the nature of the parameter k beyond this experiment, particularly how its value changes with tanks and holes of different sizes. We are not sure that $n = 1/2$ will work for all cases where we have flow from a vessel caused by the liquid head. The further information that we need can be obtained empirically if a series of experiments can be designed to test all parameters that may be of importance. Some further analysis, undertaken in the next section and particularly in Appendix 2D, can greatly simplify and direct this experimental design.

2.8 Characteristic Time

In our study of the comparison between the proposed models and actual experimental results for the tank-draining problem we have observed a phenomenon that is of great significance and bears further investigation. Recall that we mentioned previously that the challenge in model validation is to ensure that the comparison is one that truly tests the model. We have seen that the data near $t = 0$ are adequately represented by Equation 2.17 with $n = 0$ (the straight line), $n = 1$ (the exponential, Eq. 2.14), and $n = 1/2$. Clearly the data for short time do not test the model!

The use of some elementary calculus explains why any value of n seems to be adequate for short times. Consider the power-law expression in Equation 2.17b, which includes the exponential when $n = 1$ and the straight line when $n = 0$. The function $(1 + x)^\alpha$ has a series representation

$$[1 + x]^\alpha = 1 + \alpha x + \frac{\alpha[\alpha - 1]}{2} x^2 + \cdots.$$

Denoting $x = -k[1 - n]t/Ah_0^{1-n}$ and $\alpha = 1/(1 - n)$, we can then write Equation 2.17b as

$$h(t) = h_0 \left\{ 1 - \frac{kt}{Ah_0^{1-n}} + \frac{n}{2} \left[\frac{kt}{Ah_0^{1-n}} \right]^2 - \cdots \right\}.$$

When t is small, the quadratic, cubic, and higher terms will be much smaller than the first two (when $x = 0.1$, $x^2 = 0.01$, $x^3 = 0.001$, etc.), so the power-law model predicts short-time behavior closely approximated by

$$h(t) \simeq h_0 \left[1 - \frac{kt}{Ah_0^{1-n}} \right]. \qquad (2.19)$$

Two conclusions stand out. All models investigated give an identical (linear) functional dependence between h and t for small time. Hence, *a poorly designed experiment that took insufficient data would show no differences between models* and might be used to justify the assumption that q_e is independent of h. The use of that model for predictive purposes might lead to gross errors in a problem where the level changes are a significant fraction of the total height. On the other hand, we might conceive of applications where only the behavior over a short time is required. In such a situation the simplest model suffices and no further sophistication is required. This demonstrates rather forcefully the role that problem objective plays in model formulation and comparison, as shown in the logic diagram (Figure 2.1).

One further observation is in order. We have been discussing "short time" and "long time" as though time, as measured by a clock, were the pertinent variable. This is not the case, a fact that has important physical implications. Examining Equation 2.17b we see that $h(t)$ is a function only of the quantity kt/Ah_0^{1-n}. It is this quantity that must be small if the approximation used in deriving Equation 2.19 is to be valid. Just what we mean by "small" depends on the particular situation. In the example we have been discussing, a value of k/Ah_0^{1-n} equal to $0.015\ \mathrm{s}^{-1}$ was small enough to produce agreement between all three values of n at times up to 17 sec. If we could be satisfied with agreement on the order of 10 percent, the terms could be larger and still fall into the "small" category. We thus see that it is not time that must be small but a ratio t/θ, where we will call θ a *characteristic time*. The characteristic time for each model considered here is Ah_0^{1-n}/k for $0 \le n \le 1$. That θ truly has dimensions of time is easily verified: kh_0^n is the initial flow rate, with dimensions of volume per time, while Ah_o is the initial volume; a volume divided by a volume per time has dimensions of time. Hence, t/θ is dimensionless.

Realizing that t/θ is the key grouping of parameters in this problem enables us to decide on the model complexity required in any given physical situation. If, for instance, we were designing a control system to maintain a constant level in a tank when flows to and from the tank change from their desired values over a time duration t_D, then we could use the simplest model, which assumes the flow rate is independent of height ($n = 0$), if t_D/θ for the tank in question were small enough that there would be negligible error in using the simple model. On the other hand, if the deviations between the predictions of the two models should be significant, then we would have to make a decision either to obtain accurate prediction with a more complex model or to compromise with less accurate prediction and a simpler model. The decision is not crucial in this case, of course, since both models are really quite simple. The best level of compromise is often difficult to reach in more complicated situations. A careful analysis of how the model equation fits into the overall problem must always be carried out. If it is a part of a more complex mathematical description, simplicity is obviously important, whereas if the particular model is to be used alone, simplicity may not be as desirable as accurate prediction.

2.9 Scaling and Dimensions

The dimensions of the variables that appear in a problem, and the units used for measurement, are vital considerations in any scientific and engineering analysis. The requirements of dimensional consistency in equations and the use of dimensional analysis, in particular, are powerful tools. Chemical engineering curricula differ on the placement of these topics. Systems of units and dimensional consistency are often covered in courses in chemistry or physics. Dimensional analysis is often covered as a topic in one of the transport courses. We have included these topics as a series of brief appendices to this chapter so as not to break the flow for those who have studied them previously or will put dimensional analysis off to a subsequent course. We call particular attention to the treatment of orifice flow by dimensional analysis in Appendix 2D, which leads directly and naturally to the $n = 1/2$ relation.

It is useful at this time to consider some simple implications of the dimensions of the variables in the draining tank. We start by recalling that the physical relation we need is really between flow rate q_e and pressure change Δp; h entered directly only because we know that the pressure change from the liquid inflow to the orifice is proportional to liquid height, and height is the quantity that we are measuring. Now, a pressure is a force exerted per unit area, but we can also think of it as force $\times$ distance (i.e., work) divided by area $\times$ distance (i.e., volume). So pressure change is the work per unit volume, or, equivalently, the energy required per unit volume, to push the liquid through the orifice. The average velocity of the liquid going through the orifice is q_e/A_0 (flow rate is area times velocity), so the kinetic energy per unit volume is $\frac{1}{2}\rho(q_e/A_0)^2$. (Kinetic energy is mass multiplied by one-half the square of velocity. Mass is density multiplied by volume.) We know that not all of the work required to force the liquid through the orifice can be converted to kinetic energy, because the Second Law of Thermodynamics requires that some of the work be dissipated and lost. (Otherwise we could recover all of the work as kinetic energy, use the kinetic energy to do work, and build a perpetual motion machine.) We nevertheless expect the kinetic energy of the liquid and the work required to cause the liquid to flow to be of comparable magnitude, so we write

$$\rho\left[q_e/A_0\right]^2 \sim \Delta p$$

where the symbol $\sim$ means "is of the same order as." Since $\Delta p = \rho g h$, where g is the gravitational acceleration, we expect

$$q_e/A_0 \sim [gh]^{1/2}, \tag{2.20}$$

which is the $n = 1/2$ dependence found in Section 2.7. We now have the additional information, however, that we expect the constant k to be proportional to $A_0 g^{1/2}$, which permits us to test the relation by using orifices of different diameters.

The simple scaling analysis that we used here, which is of a type frequently used by engineers and physical scientists, is, in fact, a rough sketch of the derivation of the Bernoulli equation mentioned earlier. The critical assumption in that derivation is that viscous losses because of fluid friction are small compared to the work and kinetic energy contributions.

2.10 Concluding Remarks

Flow in and out of a tank illustrates the basic elements of the analysis process, including the limitations of the use of the conservation equations, the need for situation-specific constitutive relations, and the types of issues that will arise in model validation. The level controller is an elementary example of an engineering design problem. In Chapter 4 we will consider systems in which there is more than one component; this will reinforce and expand on the points made here, as well as provide an opportunity to consider problems of much greater significance and interest.

Bibliographical Notes

Model validation is addressed briefly in Chapter 14 of[*]

Denn, M. M., *Process Modeling*, Longman, London and Wiley, New York, 1986.

The literature on validation usually addresses scientific theories, not models, but the issues are the same. This is an important component of the philosophy of science, which is an area of study that is very valuable for any scientist or engineer. Some basic texts are

Hesse, M. B., *Models and Analogies in Science*, U. Notre Dame Press, South Bend, IN, 1966.
Hesse, M. B., *The Structure of Scientific Inference*, U. California Press, Berkeley, 1974.
Popper, K. R., *The Logic of Scientific Discovery*, Harper and Row, New York, 1968.

These are timeless books that address the most fundamental issues of how scientists work and how science progresses. Popper's concept of the *degree of falsifiability*, which is based on quantifying the notion that a scientific theory can never be proven, only shown to be wrong, is an extremely useful, albeit controversial, concept.

Textbooks on automatic control are designed for students who have completed at least a course in differential equations, but there is an excellent introduction to the concept of feedback and its early history for the lay reader in the first part of a collection of *Scientific American* articles, published in book form in 1955:

Automatic Control, Simon and Schuster, New York, 1955.

The book is out of print but widely available. There is also an excellent history in

Bissell, C. C., "A History of Automatic Control," in Y. Nof, editor, *Springer Handbook of Automatic Control*, Springer, New York, 2009, Chapter 4.

An accessible article in a major journal describing the development of the discipline during the critical period 1940–1960 is

Bennett, S., "The emergence of a discipline: Automatic control 1940–1960," *Automatica*, **12**, 113–121 (1976).

PROBLEMS

2.1. A wedge-shaped tank (Figure 2P.1) is filled and emptied at constant flow rates q_f and q_e, respectively. The liquid density is a constant at all times. The liquid height

---

[*] There is an obvious error at the bottom of p. 270, where the phrases in the minor premise and conclusion of the syllogism should be reversed.

is h_0 at $t = 0$. Find the height at any time. (Hint: Write the model equation in terms of h^2 in order to obtain a form that is easy to integrate.)

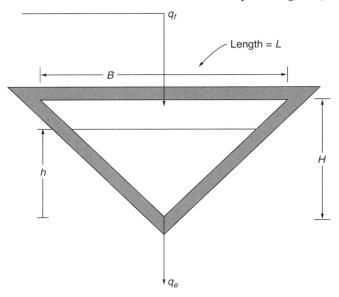

Figure 2P.1. Wedge-shaped tank filling at constant volumetric flow rate q_f and emptying at constant volumetric flow rate q_e.

2.2. The system shown in Figure 2P.2 might be used to load ore onto barges for shipment. Let W_1 be the constant mass flow rate of ore onto the barge, and let M be the mass of ore at any time. After the pile has reached a mass M_0, ore is lost over the side at a mass flow rate W_2. The loss rate is roughly proportional to the amount of ore in excess of M_0 at any time.

    a.   The barge is initially empty. Calculate the time t_0 at which the ore on the barge reaches M_0.

    b.   How does M depend on t for $M > M_0$? What happens as $t \to \infty$?

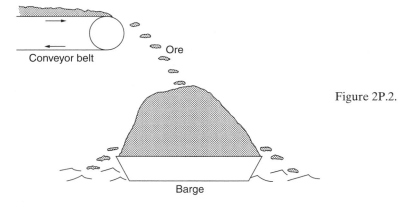

Figure 2P.2.

2.3. Consider Equation 2.6, and now suppose that we wish to use a combination of feedback and feedforward control in the form $q_e = q^* + K_{fb}(h - h^*) + K_{ff}Q(t)$.

a. Show that you obtain perfect control ($h = h^*$ at all times) for $K_{ff} = 1$. Show that steady-state offset is reduced for $K_{ff} < 1$. (Why should K_{ff} never be greater than unity?)

b. Now suppose that some input disturbances are unmeasured, or there is unknown measurement error, in which case $Q(t) = Q_m(t) + Q_u(t)$, where the subscripts m and u refer to measured and unmeasured, respectively. The controller then has the form $q_e = q^* + K_{fb}(h - h^*) + K_{ff}Q_m(t)$. Obtain the equation for the response of the controlled system. How should K_{fb} be chosen if $K_{ff} = 1$? If $K_{ff} < 1$?

c. In your physics course you may have studied the damped harmonic oscillator. If so, consider the case in which the feedback controller is *proportional plus integral*, or PI, control:

$$q_e = q^* + K_{fb}(h - h^*) + K_I \int_0^t [h(\tau) - h^*]d\tau.$$

Show that the equation for the flow in the tank becomes identical to the equation for the damped harmonic oscillator, and that steady-state offset is eliminated for $Q(t) = $ constant.

2.4. The following data were obtained in a cylindrical vessel of diameter 2.54 cm (1.0 in.) that was allowed to drain through an orifice of diameter 0.109 cm (0.043 in.).

Height (cm)	(in.)*	Time (s)
38.1	15	0
35.6	14	6.0
33.0	13	12.2
30.5	12	18.7
27.9	11	25.5
25.4	10	32.7
22.9	9	40.3
20.3	8	48.3
17.8	7	56.7
15.2	6	66.1
12.7	5	76.2
10.2	4	87.6
7.6	3	101.0
5.1	2	117.5
2.5	1	140.7

*The data were originally measured in inches and converted to centimeters.

a. Fit the equation $q = kh$ to the first five data points and estimate the error at 140.7 seconds.

b. Suppose that $q = CA_o\sqrt{gh}$ (cf. Section 2.9 and Appendix 2D). Find C. How does this value compare with the value calculated from the data in Table 2.1?

2.5. A chemical operator is draining a tank originally full of crude oil. The tank measures 0.91 m (3 ft) in diameter. The operator opens a valve with a 3.18 cm (1–1/4 in.) diameter in the tank base. He records the following data as the level falls from the initial height to 1.83 m (6 ft):

Height		
(m)	(feet)	Time (s)
3.05	10	0
2.74	9	54.0
2.29	7.5	142.9
2.13	7	172.5
1.83	6	240.2

    a.   The operator assumes that the flow rate is steady and puts a straight line through these data. When would he expect the tank to empty?

    b.   You believe that $q = kh^{1/2}$. Using these data, find k and estimate the time for the tank to empty. How different is your estimated time from that in part a?

    c.   Assume for the result in part b that $k = C A_0 \sqrt{g}$. What value do you compute for C?

2.6. A tank is filled at a constant flow rate q_f, while it empties through a valve for which the flow rate is equal to $kh^{1/2}$. Will this system come to a steady state ($dV/dt = 0$)? If so, what is the height at steady state?

2.7. A water clock is a vessel that has a shape such that the liquid level h decreases at a constant rate. In this way, water level markings can be easily calibrated with elapsed time. Water clocks were used in antiquity. Determine the shape of an axisymmetric water clock, assuming that water flows out through an orifice at the bottom at a rate proportional to $h^{1/2}$.

PROBLEMS ON DIMENSIONAL ANALYSIS (APPENDIX 2D)

2D.1. A rock of mass m falls in a vacuum under the acceleration of gravity (g). Use dimensional analysis to find an expression for the distance s fallen in time t in terms of m, g, and t. Compare with the result that you learned in your introductory physics course.

2D.2. a. A simple model of a gas assumes that the molecules are rigid spheres moving randomly with a mean speed u. The pressure, p, depends on u; on the number of moles, n; on the molecular weight, M_w; and on the volume, V. Find an expression for p.

    b. Recall from your physics or chemistry course that the average kinetic energy in this model of a gas is proportional to the absolute temperature, T. Thus, replace $M_w u^2$ in your solution to part a with a constant multiplied by T. Show that this substitution results in the ideal gas relation, $pV = nR_g T$, where in this case R_g is an unknown constant.

2D.3. When a low molar mass liquid, like water or glycerine, flows through a long pipe of length L the volumetric flow rate Q depends on the liquid density, ρ; the pipe radius, R; the liquid viscosity, η (Pa s); and the pressure gradient, or pressure change per unit length, $\Delta p/L$.

a. What are the relevant dimensionless variables, and how does Q depend on the problem variables?

b. At low flow rates a series of experiments is carried out to measure the dependence of Q on $\Delta p/L$ while holding ρ, R, and η constant. It is found that Q is proportional to $\Delta p/L$. How will Q depend on R? How does Q depend on the density? Explain the density dependence in physical terms. (This dependence was first found between 1837 and 1840 by Hagen in Germany and Poiseuille in France, and the resulting equation is known as the Hagen-Poiseuille Law.)

Appendix 2A: Estimating an Order

In Section 2.7.3 we were faced with finding the parameter n in the two-parameter constitutive relation

$$q_e = kh^n.$$

Such "power laws" appear frequently in engineering and science. n is known as the *order* of the process. With data that are reasonably free of experimental scatter it is often possible to estimate the order within reasonably narrow bounds, and hence reduce or eliminate the trial-and-error nature of the computations. This is frequently done to determine the order of chemical reactions, for example. If we take logarithms of both sides of Equation 2.15 we obtain

$$\ln\left[-\frac{dh}{dt}\right] = \ln\frac{k}{A} + n\ln h. \tag{A2.1}$$

Thus, a plot of $\ln[-dh/dt]$ versus $\ln h$ will have slope n. However, we do not have dh/dt available. If the data are reasonably free of error and closely spaced over regions where h is changing rapidly, we can approximate $-dh/dt$ as $-\Delta h/\Delta t$, where Δh refers to the change in height over a corresponding change Δt in time. Then, approximately,

$$\ln\left[-\frac{\Delta h}{\Delta t}\right] \simeq \ln\frac{k}{A} + n\ln h. \tag{A2.2}$$

The height-time data from Table 2.1 are reproduced in Table 2A.1 using average times from the three experimental runs and plotted as the logarithm of $-\Delta h/\Delta t$ versus the logarithm of h in Figure 2A.1. Since the derivative is estimated over a range of heights the data are plotted as the mean height with a band ($|-|$) to represent the range. Although no definitive statement can be made because of the spread in the data, it is clear that the mean values in the region of least scatter are best fit with a line of slope $n = 0.5$. The height changes too rapidly and there is too much

Table 2A.1. *Liquid height versus average time for the tank-emptying experiment in order to obtain* ln $(-\Delta h/\Delta t)$ *versus* ln h. ($\Delta h = 2.54$ *reflects the fact that the original heights were measured in integral values of inches.*)

h	t	$-\Delta h$	Δt	$\ln\left(\dfrac{-\Delta h}{\Delta t}\right)$	$\ln h$
30.5	0				3.4
		2.54	5.9	−0.85	
27.9	5.9				3.3
		2.54	5.4	−0.76	
25.4	11.3				3.2
		2.54	5.9	−0.85	
22.9	17.2				3.1
		2.54	6.0	−0.86	
20.3	23.2				3.0
		2.54	6.5	−0.97	
17.8	29.7				2.85
		2.54	6.5	−0.97	
15.2	36.2				2.7
		2.54	7.3	−1.06	
12.7	43.5				2.5
		2.54	7.5	−1.09	
10.2	51.0				2.3
		2.54	9.0	−1.27	
7.6	60.0				2.0
		2.54	10.7	−1.44	
5.1	70.7				1.6
		2.54	14.0	−1.71	
2.5	84.2				0.9

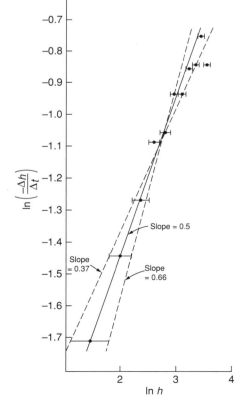

Figure 2A.1. Natural logarithm of $-\Delta h/\Delta t$ versus natural logarithm of height. The band represents the range of heights over which the slope was calculated.

scatter in these estimates of derivatives to be of any use when h is greater than 20 cm ($\ln h > 3$), and any slope will fit the data equally well. This is equivalent to our previous observation that all models are equivalent at short times.

It is essential to emphasize that, although approximating derivatives is a helpful and sensitive way to estimate order (power) when the data are good and closely spaced, any attempt to estimate k by means of the intercept will give extremely poor accuracy. For the data shown here, errors in k of the order of 20 percent can be expected because of the spread about the mean at each point. The system equation must be integrated to obtain the parameter k.

Appendix 2B: Systems of Units

There are five systems of units that are commonly used in English-speaking countries, two "metric" and three "Imperial." It is necessary to be familiar with each. The quantities we generally measure are length (L), time (θ), mass (M), force (F), and temperature (T). The fundamental unit of time is always seconds, but the units of one or more of the other quantities vary from system to system. A basic relation necessary to our discussion is that between force and mass, *Newton's principle of acceleration (Newton's Second Principle),* which states that the force required to accelerate a given mass uniformly is equal to the product of mass and acceleration:

$$f = ma. \tag{2B.1}$$

The dimensions of acceleration are length per time per time or, symbolically,

$$a[=]L\theta^{-2}. \tag{2B.2}$$

The symbol $[=]$ means "has dimensions of" or "has units of," depending on the context. Thus, the dimensions of force,

$$F[=]ML\theta^{-2}, \tag{2B.3}$$

are defined in terms of M, L, and θ.

The system of units most commonly used in scientific work through most of the twentieth century is the cgs, or *centimeter-gram-second,* system. Here, length is measured in centimeters (cm), mass in grams (g), time in seconds (s), and temperature in degrees centigrade or Celsius (°C). From Equation 2B.1, then,

$$\text{force}[=] \text{g cm s}^{-2}.$$

The cgs unit of force is called the dyne (d), so we have the definition

$$\text{d}[=] \text{g cm s}^{-2}.$$

Now there is an important, but subtle, point here. We have introduced the units centimeter, gram, second, and dyne, *four* quantities of which only *three* are independent. If we wish to measure force in dynes, mass in grams, length in centimeters, and time in seconds, then Equation 2B.1 does not have consistent units, and we need to

Table 2B.1. *Units of the cgs*
(centimeter-gram-second) system

Length	[=] centimeter, cm
Mass	[=] gram, g
Time	[=] second, s
Force	[=] dyne, d
Temperature	[=] degrees centigrade, °C
g_c	= 1 g cm/d s^2

rewrite it to include a conversion factor, commonly called g_c:

$$g_c f = ma \tag{2B.4}$$

$$g_c [=] \frac{\text{g cm}^2}{\text{d s}^2}.$$

The conversion factor is frequently omitted in the cgs system because its numerical value is unity. Such is not the case in all systems, however, and it is good practice to include g_c whenever force and mass both appear. Finally, let us recall that the *weight* of an object, w, is the force caused by the acceleration due to gravity,

$$g_c w = mg. \tag{2B.5}$$

The acceleration due to gravity is approximately 980 cm/s^2, so the weight of a 1 g mass is 980 d. There is an unfortunate confusion between mass and weight. The mass of an object is the same anywhere in the universe; the weight of an object depends on the gravitational acceleration and varies slightly from place to place, even on earth. The cgs system is summarized in Table 2B.1.

A closely related system is the S.I., or *International System* (Système International d'Unités), in which the units of length, mass, time, and temperature are, respectively, meter (m), kilogram (kg), second, and Kelvin (K). (A change of 1 K is the same as a change of 1°C, but the Kelvin scale starts at absolute zero; i.e., 0°C = 273.15 K.) The unit of force, called a Newton (N), is numerically equal to one kilogram meter per second per second, so that g_c again has a numerical value of unity. The system is summarized in Table 2B.2. The S.I. has been adopted by most countries and is now the required system of units for the *AIChE Journal,* published

Table 2B.2. *Units of the S.I. (International System)*

Length	[=]	meter, m	1 m = 100 cm
Mass	[=]	kilogram, kg	1 kg = 1,000 g
Time	[=]	second, s	
Force	[=]	Newton, N	1 N = 10^5 d
Temperature	[=]	Kelvin, K	1 K = 1°C
g_c	=	1 kg m/N s^2	

Table 2B.3. *Units of the fps (foot-pound-second) system*

Length	[=]	foot, ft	1 ft = 30.48 cm
Mass	[=]	pound-mass, lb_m	1 lb_m = 453.59 g
Time	[=]	second, s	
Force	[=]	poundal, lbl	1 lbl = 13,826 d
Temperature	[=]	degrees Fahrenheit, °F	1°F = 5/9°C
g_c	=	1 lb_m ft/lbl s^2	

by the American Institute of Chemical Engineers, and most other technical publications. It has not been widely adopted by U.S. engineers, however, so it remains necessary to be familiar with the various systems and their interconversions.

The basic Imperial system is the fps, or *foot-pound-second* system. Here the units of length, mass, time, and temperature are, respectively, the foot (ft), pound-mass (lb_m), second, and degree Fahrenheit (°F). In the fps system the unit of force is defined in such a way as to keep g_c numerically equal to unity. This force unit, the poundal (lbl), is defined as

$$1 \text{ lbl} = 1 \text{ lb}_m \text{ ft s}^{-2}.$$

The system is summarized in Table 2B.3. The fps is a logical system, but it is seldom used because of the unfortunate general confusion between mass and weight. Although incorrect, it has become common to talk of both mass and force, particularly the force, weight, in pounds, as though the two were the same.

The Imperial engineering system, sometimes simply called the Imperial system, was developed in an attempt to avoid confusion, and it is the most common system in use in the United States. In this system, summarized in Table 2B.4, the units of length, mass, time, and temperature remain the foot, pound-mass, second, and degree Fahrenheit. A new unit of force is defined, however, the *pound-force* (lb_f), with

$$1 \text{ lb}_f = 32.174 \text{ lbl}$$

or, from Equation 2B.4,

$$1 \text{ lb}_f = 32.174 \text{ lb}_m \text{ ft/s}^2$$

$$g_c = 32.174 \text{ lb}_m \text{ ft/lb}_f \text{ s}^2.$$

Table 2B.4. *Units of the Imperial system*

Length	[=]	foot, ft	1 ft = 30.48 cm
Mass	[=]	pound-mass, lb_m	1 lb_m = 453.59 g
Time	[=]	second, s	
Force	[=]	pound-force, lb_f	1 lb_f = 4.45 × 10^5 d
Temperature	[=]	degrees Fahrenheit, °F	1°F = 5/9 °C
g_c	=	32.174 lb_m ft/lb_f s^2	

The choice of 32.174 is, of course, not arbitrary. The acceleration due to gravity in Imperial units is 32.174 ft/sec² at sea level and 45° latitude, so that when we express the *weight* from Equation 2B.5 as

$$w = \frac{g}{g_c} m$$

we find that the weight of one pound-mass is numerically equal to one pound-force. Agreement with common usage is gained at the expense of a conversion factor that is not unity and much opportunity for error in making conversions between the different systems. Furthermore, it is important to note that the Imperial engineering system is an earthbound system; g_c is a fixed number, whereas g varies slightly from point to point on earth, but to sufficient accuracy the numerical ratio g/g_c is always unity. In a markedly different gravitational field, however, the ratio g/g_c will differ greatly from unity and the sole advantage of the Imperial engineering system, the numerical equality of mass and weight, is lost.

Finally, we note in passing that there is another Imperial system that is used on occasion, the *gravitational system*. The units of length, mass, time, force, and temperature are, respectively, foot, slug, second, pound-force, and degree Fahrenheit. The slug is defined as

$$1 \text{ slug} = 32.174 \text{ lb}_m$$

$$g_c = 1 \text{ slug ft/lb}_f \text{ s}^2.$$

There are also two metric systems using a gram-force and kilogram-force. These systems make no sense at all.

A few words are in order about temperature. The zero points on the centigrade and Fahrenheit scales are completely arbitrary (0°C = freezing point of water, 0°F = freezing point of water less 32 degrees!) and the conversion factors 1°F = 5/9°C and 1 K = 1°C refer only to temperature *differences*. The more natural scales are the *absolute* Rankine (°R) and Kelvin (K) scales, where the zero corresponds to the point at which molecular motion in an ideal gas ceases. The relations are

$$T(^\circ R) = T(^\circ F) + 459.58,$$

$$T(K) = T(^\circ C) + 273.15.$$

It is necessary to distinguish carefully between the common and absolute scales. Fundamental relations in physics and chemistry require absolute temperatures.

Appendix 2C: Common Units

Comprehensive tabulations of physical property data are available in references such as the *International Critical Tables,* the *Handbook of Chemistry and Physics,* and the *Chemical Engineers' Handbook.* Extensive physical property databases are also part of commercially available flowsheet simulation software, and databases from the U.S. National Institute of Standards and Technology (NIST) are available electronically at the time of this writing at http://webbook.nist.gov.

Densities are often tabulated with the units of g/cm^3, and conversion to kg/m^3 simply requires multiplication by 1,000. Multiplication by 62.4 converts the tabulated values from g/cm^3 to lb_m/ft^3, still the most commonly used unit in the United States. Gas densities for many situations (low pressures and high temperatures relative to the critical values) can be estimated using the ideal gas "law,"

$$\rho_{\text{gas}} = M_w \left(\frac{p}{R_g T} \right),$$

where M_w is the molecular weight and $R_g = 82.06$ for p in atm, T in K, n in g-mol (1 atm $\cong$ 14.7 lb_f/ft^2 = 10^5 N/m^2 = 10^4 d/cm^2). For example, the average molecular weight of air is approximately 29, so the density of air at 1 atm and 0°C (273 K) is

$$\rho_{\text{air}} = 29 \left(\frac{1}{82.06 \times 273} \right) = 1.29 \times 10^{-3} \text{ g/cm}^3.$$

In the SI system, the density of air at atmospheric pressure at 273 K (0°C) is 1.29; in general, the densities of gases are approximately unity in the SI system. The density of water is about 1,000 in the SI system.

The concentration, or mass per unit volume of a particular species, is not as easily measured as density, because the method of measurement must be specific to the substance of interest. The development of sensors is an active area of research in chemical engineering. Concentration can be expressed in mass/volume or moles/volume, and common units and conversions are

$$g/\text{liter} = kg/m^3 = g/cm^3 \times 1,000$$
$$lb_m/ft^3 = g/cm^3 \times 62.4$$
$$g\text{-mol}/cm^3 = g/cm^3 / M_w$$
$$lb\text{-mol}/ft^3 = lb_m/ft^3 / M_w$$

Very low concentrations are often expressed in mg (10^{-3} g) and μg (10^{-6} g) per unit volume, usually m^3. Expressing component quantities in either mass or mole fraction gives a dimensionless ratio and avoids unit conversions:

$$\text{Mass fraction of } A = \frac{\text{Mass of } A}{\text{Total mass}}$$

$$\text{Mole fraction of } A = \frac{\text{Moles of } A}{\text{Total moles}}.$$

It has become common in some applications to use ppm (parts per million) when measuring and controlling very low concentrations, such as impurities in electronic materials or pollutants in the air or water. Parts per million is a mass fraction for solids and liquids and a mole fraction for gases. Some useful conversions are:

Mass fraction of A (liquid or solid) $\times$ 10^6 = ppm of A

Mass fraction of A (liquid or solid) $\times$ [density of mixture] = concentration of A, mass/volume

Percent of A (liquid or solid) $\times 10^{-4}$ = ppm of A
ppm of A (gas at 25°C and 1 atm) $\times M_W/24.5$ = mg/m³

Energy and work have the same dimensions, force $\times$ distance. The SI unit of energy, the joule (1 J = 1 N $\times$ 1 m = 1 kg m²/s²), is becoming the standard, but the calorie and British Thermal Unit (BTU) remain in widespread use in engineering practice. Power is work per unit time, and the common units are the watt (1 J/s) and the horsepower. Conversion factors are given below:

$$\text{Joule} \times [0.738] = \text{ft lb}_f$$
$$\text{Joule} \times [0.239] = \text{calorie, cal}$$
$$\text{ft lb}_f \times [0.001286] = \text{BTU (British Thermal Unit)}$$
$$\text{calorie} \times [1,000] = \text{kilocalorie, Kcal (or Calorie, C)}$$
$$\text{Joule} \times [0.000948] = \text{BTU}$$
$$\text{Joule} \times [3.6 \times 10^6] = \text{kWh (kilowatt hour)}$$
$$\text{kWh} \times [3412] = \text{BTU}$$
$$\text{BTU/sec} \times [1.414] = \text{HP}$$

The kilocalorie (Calorie with an uppercase "C") is still commonly used in the laboratory and by those watching their diets or interested in exercise. Food energy contents are tabulated in kilocalories, but people are very careless about the prefix or the capitalization of C, and the term *calorie* is often used. Food energy content ranges from 25 Kcal for a half cup of carrots to 165 Kcal for a half cup of ice cream. A person in a developed country uses between 2,000 and 2,500 Kcal per day, making the average power output equal to about 100 watts. A well-trained athlete can produce between 1,000 and 2,000 W for periods on the order of hours. The total power consumption per person in the United States is about 10,000 W, or 10 kW. This is equivalent to a yearly energy use for the United States of about 89 exajoules (10^{18} J) or 85 Quads (10^{15} BTU). It is also common to express energy use in terms of the important fuels as follows:

$$1 \text{ 42-gallon barrel of crude oil} = 5.8 \times 10^6 \text{ BTU}$$
$$1 \text{ ft}^3 \text{ of natural gas} = 1,013 \text{ BTU}$$
$$1 \text{ ton (2,000 lb}_f\text{) of coal} = 25 \times 10^6 \text{ BTU}$$

Process engineering calculations are usually done in the United States using the BTU, although there is a trend developing to using joules. There is usually a considerable amount of unit conversion required in process energy balances, since properties such as heat capacity and heat of reaction are frequently found in older sources in cal/g °C and cal/g-mol, respectively.

Although we will not deal with applications of the law of conservation of momentum in this text, it is useful to know some of the characterizing variables and units. Shear stress is a force (Newton) per unit area, and has units of N/m² or pascal (Pa).

Shear rate is a velocity divided by distance, and has units of 1/time. A plot of experimental data of shear stress versus shear rate is often linear, and leads to a definition of viscosity as the ratio of shear stress to shear rate. Viscosity is thus measured in Pascal seconds; the cgs unit, the poise (p), is also used (1 Pa·s = 10 p; 1 mPa = 1 cp). The viscosity of water is about 1 cp or 1 mPa·s at room temperature, and the viscosity of air is about two orders of magnitude less than that of water.

Appendix 2D: Dimensional Consistency and Dimensional Analysis

Equations are dimensional; every term in the rate form of the equation of conservation of mass, for example, has dimensions of mass/time, in units of kg/s, lb_m/s, or some equivalent system. It is essential to check that every term in the equation does indeed have the same dimensions; this is a seemingly trivial step, but failure to ensure dimensional consistency is probably the most common source of error in the formulation of mathematical models. (The second is probably failure to carry out unit conversions correctly.)

As we saw in the discussion of characteristic times, it is sometimes useful to put terms in dimensionless form. It is always possible to formulate a dimensional equation in dimensionless form. There are often conceptual advantages in a dimensionless formulation, one of which is to reduce the number of independent parameters that occur in the model equations. Another is the opportunity to invoke *dimensional analysis*, a type of analysis formalized by the physicist Percy Bridgman that can lead to powerful insights into data analysis and the form of constitutive equations.

Dimensional analysis is based on the *Buckingham Pi Theorem*, which is a straightforward consequence of theorems in linear algebra:

> *Let the number of dimensional variables describing a system equal* V, *and let the number of dimensions equal* D. *Then the description of the system can be expressed in terms of* G = V − D *independent dimensionless variables made up of combinations of the* V *dimensional variables.*

(Rigorously, G is the maximum number of dimensionless groups, but the rare exceptions are not important to our discussion.) There may also be dimensionless parameters in the basic description, and these are in addition to the number cited here. There are formal ways of generating the dimensionless groups, but most people find them by inspection. We will illustrate the process here with the example of the draining tank.

The characterizing variables in the tank-draining problem, with their dimensions, are ρ [M/L^3], g[L/θ^2], h [L], q [L^3/θ], and A_0 [L^2]. There are five variables and three dimensions, so we expect two independent dimensionless groups in any description of the process. We see immediately that the density is the only variable with units of mass, so there is no way in which density can be combined with any other variable to form a dimensionless quantity. Hence, density cannot possibly be one of the variables describing the process. (We will return to this point subsequently.) That leaves g, h, q, and A_0, with dimensions of length and time: four variables and two dimensions, hence two groups. It is easy to construct these groups by inspection. The only variables that include time are g and q, so clearly the ratio q^2/g must be part of one of the groups. The remaining length dimensions are balanced by including $A_0^2 h$, giving the dimensionless group $q^2/A_0^2 gh$. An independent dimensionless group is clearly A_0/h^2. Hence, the dimensionless description of the tank draining must be of the form

$$q^2/A_0^2 gh = f(A_0/h^2), \qquad (2D.1)$$

where f is a function whose form we do not know.

This is as far as dimensional analysis can take us, and it is not quite far enough, since we do not have a relation that can be solved for q in terms of h. We can apply some physical insight to this particular problem at this point, however, and go further. We are only interested in cases in which $A_0 \ll h^2$ (the diameter of the hole is very small relative to the height of liquid for most of the time of interest), hence it is likely that the limiting case $A_0/h^2 \to 0$ is the only one of interest. In that case, we are interested only in the value of the function f when its argument is zero, which is a constant. Thus, we can conclude that $q^2/A_0^2 gh$ must equal a constant, which we will denote C; i.e.,

$$q = CA_0\sqrt{gh}. \qquad (2D.2)$$

Hence, we recover the square-root dependence, but with the important added information about the functional form of the coefficient.

Now, let us return to the issue of the density. We have ignored any dissipative losses associated with the flow through the orifice. (In thermodynamic terms, we have assumed that the process is reversible and we have assumed that the losses required by the second law of thermodynamics are not important.) This is the same assumption as made in Section 2.9. The losses depend on the viscosity; had we included the viscosity we would have obtained another dimensionless group, known as the Reynolds number, $q\rho/A_0^{1/2}\eta$, where η is the viscosity. Our result is the correct one for the limit of very high Reynolds number; in this limit, C is a constant equal to about 0.6. Dimensional analysis can only give a relation between the variables that are assumed to be important. If an important variable is not included on the list, it cannot appear in the result.

Appendix 2E: Least-Squares Fitting

It is often necessary to pass the "best" line through a data set; we did so in Section 2.7.3, for example. The most common procedure is to use the *method of least squares*, in which the line is chosen so as to minimize the sum of the squares of the differences between the line and the data points. Least-squares fitting provides an explicit relation for the coefficients, which is not the case with other procedures (minimizing the sum of the absolute values of the differences, for example, or minimizing the maximum deviation). Least-squares fitting also has a firm theoretical foundation as the optimal procedure for cases in which the experimental errors in the data have certain common random properties. These topics are covered in specialized texts and are beyond the scope of our discussion here, which is to present the method for use.

Let the set of numbers $\{y_i\}$, $i = 1, 2, \ldots, N$ be the N values of the variable y measured at the N values $\{x_i\}$, $i = 1, 2, \ldots, N$ of the independent variable, x. We seek the coefficients a and b in the equation

$$y = ax + b. \tag{2E.1}$$

We construct the sum of the squares of the deviations,

$$E = \tfrac{1}{2} \sum_{i=1}^{N} (y_i - ax_i - b)^2. \tag{2E.2}$$

E is minimized by setting the derivatives $\partial E/\partial a$ and $\partial E/\partial b$ to zero. The result is a pair of linear equations for a and b,

$$\left(\sum_i x_i^2 \right) a + \left(\sum_i x_i \right) b = \sum_i x_i y_i, \tag{2E.3a}$$

$$\left(\sum_i x_i \right) a + Nb = \sum_i y_i. \tag{2E.3b}$$

The coefficients are simply sums of the experimental quantities. These equations can be solved to get explicit equations for the two parameters,

$$a = \frac{N \sum x_i y_i - \sum x_i \sum y_i}{N \sum x_i^2 - (\sum x_i)^2}, \tag{2E.4a}$$

$$b = \frac{\sum x_i^2 \sum y_i - \sum x_i y_i \sum x_i}{N \sum x_i^2 - (\sum x_i)^2}. \tag{2E.4b}$$

These are the equations used to calculate the slopes and intercepts of the lines in Figure 2.11, where y is the square root of the height and x is the time. Most graphical software includes subroutines for least-squares fitting.

The Balance Equation

3.1 Introduction

In Chapter 2 we introduced the concept of a balance equation to account for the total mass in the control volume. Mass is a conserved quantity that is neither created nor destroyed, so the concept of a balance equation is straightforward. We can and often do write balance equations for quantities that are *not* conserved, and it is appropriate to digress briefly to consider this point. One important example of a quantity that is not conserved is the mass of a reactive species. Suppose, for example, we wish to model the distribution and metabolism of the anticancer drug methotrexate in the human body. Methotrexate is not conserved: It enters the body and then disappears because of metabolism. Nevertheless, we are able to write a balance equation for this chemical species. (Note that the number of *atoms* of each of the elements making up the drug *is* a conserved quantity.)

Most people gain experience in the use of balances through personal finance. Wealth, be it personal, national, or global, is not a conserved quantity. Nevertheless, we can and do account for wealth, typically through balancing a checkbook or analyzing monthly statements from the bank. In this chapter we will illustrate the application of balance equations to the problem of determining the true cost of future expenditures. This is a problem of inherent interest to most of the population, but the *net present worth* accounting principle outlined here is of particular relevance to engineers involved in project planning and design.

3.2 Net Present Worth

Let us suppose we are about to construct a manufacturing plant. The total cost of the project includes the cost of construction, plus the anticipated cost of operation and maintenance (O&M) over the expected lifetime of the plant. One way of comparing the costs of competing projects is to determine how much capital we would need in hand today to pay all future costs; this is called net present worth accounting.

Let $P(t)$ be the principal (i.e., total capital) available at any time, and $C(t)$ the rate at which costs are incurred at any time; i_d is the interest rate received on our invested capital, and I_d is the rate of inflation, which increases our costs. We will assume that the unit of time is one day, so the subscript d on the interest and inflation rates denotes "daily"; to obtain annual rates we multiply these rates by 365 days. By using a daily rate of interest and inflation we are assuming that interest on our capital is compounded daily and that costs are adjusted daily for inflation; daily adjustment is unrealistic, but it enables us to think of the financial process as occurring continuously in time.

Principal increases because of interest from investment, and it decreases because of outlays for O&M expenses. The rate of change of capital equals the rate of increase (interest rate multiplied by principal) less the rate of payment of expenses; that is,

$$\frac{dP}{dt} = i_d P - C. \tag{3.1}$$

Our costs increase at a rate proportional to the inflation rate:

$$\frac{dC}{dt} = I_d C. \tag{3.2}$$

We could explore various scenarios in which the interest and inflation rates vary with time. For simplicity of this analysis we assume that i_d and I_d remain unchanged over the lifetime of the project. The more general treatment for arbitrary time dependence of the rates requires only a few additional steps in the calculus.

Let C_{d0} be the daily rate of O&M expenditures at the start of the project ($t = 0$); then the solution to Equation 3.2 is

$$C(t) = C_{d0} e^{I_d t} \tag{3.3}$$

and the balance equation for principal becomes

$$\frac{dP}{dt} = i_d P - C_{d0} e^{I_d t}. \tag{3.4}$$

The starting principal at $t = 0$ is P_0.

Equation 3.4 has a form that we have already seen in Section 2.6.2, and we again use the integrating factor to rewrite the equation in the form

$$\frac{d}{dt} \left[e^{-i_d t} P \right] = -C_{d0} e^{-i_d t} e^{I_d t}. \tag{3.5}$$

Integrating both sides from $t = 0$ to any time t, and using the fact that $P = P_0$ at $t = 0$, gives

$$P(t) = P_0 e^{i_d t} - C_{d0} e^{i_d t} \int_0^t e^{-(i_d - I_d)} d\tau,$$

or

$$P(t) = P_0 e^{i_d t} + \frac{C_{d0}}{i_d - I_d} \left[e^{I_d t} - e^{i_d t} \right]. \tag{3.6}$$

Let θ_d denote the useful life of the plant. We need to have just sufficient capital on hand at $t = 0$ to ensure that everything is spent during the plant lifetime, with the last expenditure at $t = \theta_d$. Thus, we determine P_0 by requiring that $P(\theta_d) = 0$. Setting $t = \theta_d$ in Equation 3.6 and solving for P_0 thus gives

$$P_0 = \frac{C_{d0}}{i_d - I_d} \left[1 - e^{-(i_d - I_d)\theta_d} \right]. \tag{3.7}$$

Note that only the *net* interest rate – interest rate less inflation rate – is relevant. This number is likely to stay relatively constant over long periods despite fluctuations in the individual rates. It is possible to show that Equation 3.7 is the correct result when i_d and I_d vary in time, as long as the difference remains constant.

We selected daily compounding in order to treat the financial process as continuous in time. It is more convenient to work with annual rates and annual costs. The initial cost on an annual basis is $C_0 = 365C_{d0}$. The total lifetime in years is $\theta = \theta_d/365$. Annual interest and inflation rates are, respectively, $i = 365i_d$ and $I = 365I_d$. Equation 3.7 can then be written equivalently as

$$P_0 = \frac{C_0}{i - I}\left[1 - e^{-(i-I)\theta}\right]. \tag{3.8}$$

The quantity multiplying C_0 is known as the *present worth factor*, or *PWF*, and it represents the number of years of annual O&M expenses that would be needed at the start of the project in order to pay O&M expenses over the plant lifetime:

$$PWF = \frac{1}{i - I}\left[1 - e^{-(i-I)\theta}\right]. \tag{3.9}$$

If the PWF is 10, for example, then we would need ten times the annual O&M cost in hand at the start of the project in order to obtain enough from investment to be able to pay O&M costs over the lifetime. To this we would have to add the construction capital to obtain the net present worth (or net present cost) of the project.

EXAMPLE 3.1 Let us suppose that the net interest rate, $i - I$, is 8 percent (0.08) and the plant lifetime is expected to be 20 years. Then

$$PWF = \frac{1}{0.08}\left[1 - e^{-0.08 \times 20}\right] = \frac{1 - 0.202}{0.08} = 9.98.$$

We thus obtain a PWF of just under 10, so we need ten times the annual O&M cost in hand at the start of the project.

3.3 Borrowing Money

The balance equations for money in calculating the cost of a loan are similar to those in the preceding section. Suppose we wish to borrow P_0 units of currency in order to purchase a house. The terms of the contract are that we will repay the loan at a constant rate of m_d (m is for "mortgage payment"), and we take the payment rate for the sake of the analysis to be on a daily basis; the interest rate on the outstanding balance of the loan is i_d.

The balance equation for the loan is as follows: the rate of change of outstanding principal (P) equals the rate of increase from interest charges (i_dP) less the rate of payment (m_d):

$$\frac{dP}{dt} = i_dP - m_d. \tag{3.10}$$

At $t = 0$ the outstanding principal is P_0.

Equation 3.10 is separable because m_d is a constant, and it can be rewritten as

$$\frac{dP}{i_d P - m_d} = dt,$$

which is readily integrated. Alternatively, we note that the change of variable $p = P - m_d/i_d$ leads to the equation

$$\frac{dp}{dt} = i_d p$$

with solution

$$p = p_0 e^{i_d t}$$

or, equivalently,

$$P = \frac{m_d}{i_d} + \left(P_0 - \frac{m_d}{i_d}\right) e^{i_d t}. \tag{3.11}$$

If we wish to use annual payments and annual interest rates, we may replace m_d by $m = 365 m_d$ and $i = 365 i_d$ to obtain

$$P = \frac{m}{i} + \left(P_0 - \frac{m}{i}\right) e^{i t_y}, \tag{3.12}$$

where $t_y = t/365$ is the time in years.

Home buyers frequently wish to determine the relation between the size of the loan, the mortgage payments, the interest rate, and the duration of the loan. Let θ be the number of years over which the loan is to be paid off. Then $P(\theta) = 0$ (no balance of principal after θ years) and we have

$$0 = \frac{m}{i} + \left(P_0 - \frac{m}{i}\right) e^{i\theta},$$

or, after some algebra,

$$m = \frac{i P_0}{1 - e^{-i\theta}}. \tag{3.13}$$

If $i\theta$ is of order three (e.g., $i = 0.10$ and $\theta = 30$ years), e^{-3} is only 0.05 and the denominator is close to unity, so the annual payment is within 5 percent of the interest computed on the original principal.

It is common to take a loan for a fixed period, but to pay it after a shorter time with a "balloon payment" equal to the outstanding principal. Suppose we compute the payments as in Equation 3.13 for a loan over θ years, but, in fact, wish to pay the balance after θ_p years, where $\theta_p < \theta$. By substituting Equation 3.13 into Equation 3.12 and setting $t_y = \theta_p$ we obtain, after some algebra,

$$P(\theta_p) = P_0 \frac{1 - e^{-i(\theta - \theta_p)}}{1 - e^{-i\theta}}. \tag{3.14}$$

EXAMPLE 3.2 Suppose you took out a loan at 7.5 percent to be repaid over 30 years, but wished to pay the balance at the end of 10 years. The outstanding balance after ten years, $P(10)$, is then

$$\text{balance} = P(10) = P_0 \frac{1 - e^{-0.075\,(30 - 10)}}{1 - e^{-0.075\,(30)}} = 0.867\, P_0.$$

With interest payments, you will therefore have reduced the principal by about 13 percent over the first 10 years, which comprise $33^1/_3$ percent of the 30-year period.

3.4 Concluding Remarks

The financial concepts introduced in this chapter are extremely important in their own right, and they are typically covered in considerably more detail in the chemical engineering curriculum in the "capstone" design course and, perhaps, in a course in engineering economics. Our primary objective here has been to show that the approach to writing a balance equation is quite general and can be applied to many situations that appear to be outside the scope of the focus of this chemical engineering text.

PROBLEMS

3.1. Banks often advertise continuous compounding of interest on savings accounts, by which they mean that interest is added continuously to the account at a rate equal to the product of the current account balance and the annual interest rate. (Interest rates are always advertised in percentages, so the proper rate to use for calculation is the advertised rate in percent divided by 100.) Compare the return after one year of continuous compounding at 5 percent to the return with annual, semiannual, and quarterly compounding.

3.2. Show that Equation 3.9 for the present worth factor remains valid even when i_d and I_d vary in time, as long as the difference remains constant. (Hint: For this case the integrating factor is $\exp(-\int_0^t i_d(\tau)d\tau)$.)

3.3. The following population data are available for Iceland and the Faroe Islands (lumped together, because Iceland's independence from Denmark came only in 1944):

Year	Mean Population	Births	Deaths
1921	116,000	3,215	1,708
1922	118,000	3,214	1,491
1923	119,000	3,264	1,542
1924	120,000	3,150	1,730
1925	122,000	3,153	1,457
1926	124,000	3,267	1,320
1927	126,000	3,221	1,449
1928	128,000	3,162	1,318
1929	130,000	3,219	1,490
1930	132,000	3,441	1,522

Neglecting immigration and emigration, write a balance equation on people and construct a model of the population as a function of time.

    a.   Assume that the annual numbers of births and deaths are both constant, and use average values. Compare your result with the annual population data.

  b. Assume that the numbers of births and deaths annually are constant fractions of the total population, and use average values. Compare your result with the annual population data.

  c. Compare the two results and comment.

  d. The combined population of Iceland and the Faroe Islands in 2009 was 368,000. What do your models predict?

3.4. The following are population data for the United States during the period 1930–1960:

Year	Population in Millions	Births/Thousand	Deaths/Thousand
1930	123	21.3	11.3
1940	132	19.4	10.8
1950	151	24.1	9.6
1960	179	23.7	9.5

Immigration and emigration data are not included.

  Construct a model of population growth and check against the actual values of 249 million recorded in the 1990 Census and 281 million recorded in 2000. Comment on your result.

4 Component Mass Balances

4.1 Introduction

The problems studied in Chapter 2 illustrate the use of the *overall mass balance*. Most chemical engineering applications, whether in biotechnology, chemical or materials processing, or environmental control, involve a number of distinct mass species that might or might not react chemically with one another. In this chapter we will consider mass balances for multicomponent systems in which there is a single phase (i.e., we exclude immiscible oil-water systems, solid-liquid suspensions, etc.) and the component species are nonreactive. With this foundation we can go on to the far more interesting and relevant reactive and multiphase systems in subsequent chapters.

4.2 Well-Stirred Systems

We consider again the flow system shown in Figure 2.4, but we now presume that there are two components; for specificity we will take these to be dissolved table salt (NaCl) and water, but they could be any two completely miscible, nonreacting materials. The flow diagram is shown in Figure 4.1, where to each stream we associate a density and concentration. The concentration of the salt, in mass units (e.g., kg/m^3), is denoted c, while subscripts f and e again denote feed and effluent streams, respectively. Only one mass concentration, together with the density, is required to define this binary system, since the mass concentration of water is simply $\rho - c$ (total mass/unit volume less mass of salt/unit volume). Salt concentrations are easily measured; the most elementary way is to evaporate the water from a known volume and weigh the residual salt, but a better way is to measure the electrical conductivity, which can easily be correlated with the ionic concentration.

Referring to Figure 4.1, we see that we have already chosen to characterize the mass by measuring density and concentration. Selection of the control volume for the examples in Chapter 2 was straightforward: We were interested in the total mass in the tank, so we simply used the tank as the control volume. Now we need to consider the possibility that different parts of the tank may have different concentrations of

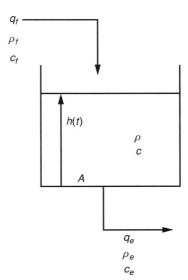

Figure 4.1. Schematic of flow of a two-component system.

salt, in which case we would be unable to characterize the salt in the entire tank by a single number. If a concentrated salt solution were to be introduced to a tank initially filled with water, for example, we would expect to find higher concentrations of salt near the inlet line than in the bottom of the tank, where the exit nozzle is located. (Consider what happens when you add milk to your coffee. The two liquids are completely miscible, but if you are careful with the addition you can get a milk-rich layer to stay above the coffee. In that case the average concentration of milk in the cup is not useful in characterizing the taste from different sips. This problem is resolved – if one desires uniformity – by stirring with a spoon.)

We must choose our control volume appropriately if we wish to deal with spatial variations in salt concentration, since the control volume must be such that we can uniquely characterize each fundamental variable. Our logic diagram for constructing a model, initially introduced as Figure 2.2, must therefore contain a step to ensure that we are using the correct control volume. Together with the discussion of constitutive equations in Chapter 2, we arrive at the flow diagram in Figure 4.2, which is adequate to deal with all problems of interest.

There are established ways to deal with spatial variation, but that is a complication that we do not have to address for this elementary problem. It is quite easy to mix waterlike liquids,[*] and we assume that the mixing is sufficiently effective to ensure complete uniformity of concentration throughout the tank. Thus, a sample drawn from one location in the tank will be identical to a sample drawn from any

---

[*] Mixing is often done by installing one or more impellers in a tank with appropriate baffles (usually four narrow strips of the same materials as the tank, mounted vertically at ninety-degree separations). Many chemical process vessels are equipped with mixers that provide power on the order of 200 to 2,000 W/m^3 (about one to ten horsepower per 1,000 gallons). Mixing can also be effected by liquid jetting into the tank at an appropriate rate and location, or by a gas sparged into the tank through a properly designed set of orifices. Very large wastewater treatment reactors, often with volumes in excess of 10^6 L (1,000 m^3, or a bit over 250,000 U.S. gallons), are mixed using air spargers.

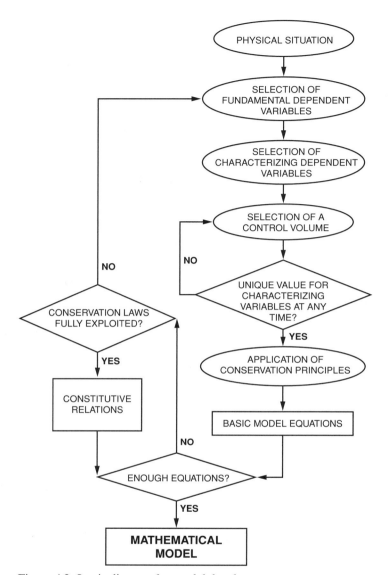

Figure 4.2. Logic diagram for model development.

other location at the same time; hence, we can denote the concentration at any time throughout the tank by a single value, denoted $c(t)$. Similarly, the density throughout the tank at any time is denoted by a single value, $\rho(t)$. We may therefore take the tank as the control volume. This *well-stirred,* or *perfectly mixed,* assumption must be verified experimentally. Experience shows that it works well in tanks where the height-to-diameter ratio is of order unity, while it fails if the ratio becomes of order ten. It works well for waterlike liquids, but it is difficult to achieve for very viscous liquids like molten polymers.

There is one further consequence of the well-stirred assumption: Since the composition and density are each the same throughout the tank at any time, it must be true that the composition and density at the exit are the same as everywhere else in the tank. Thus, there is no reason to distinguish between the values in the tank and those in the stream leaving the tank at any time, and we may drop the subscript e from the density and concentration in the effluent stream.

The principal of conservation of mass is unchanged by the fact that we now have two component species, and we still write that *the rate of change of total mass in the control volume (the tank) equals the rate at which mass enters less the rate at which mass leaves.* As before, the total mass is $\rho A h$; the rate at which mass enters is $\rho_f q_f$ (mass/unit volume × volume/time), whereas the rate at which mass leaves is ρq_e. Note that, because of the perfect-mixing assumption, we do not distinguish between ρ and ρ_e. We can thus write

$$\frac{d\rho \, Ah}{dt} = \rho_f q_f - \rho q_e. \tag{4.1}$$

We are no longer assuming that ρ and ρ_f must be the same, and ρ may be varying with time, so we cannot introduce the simplification that led to Equation 2.3.

Conservation of mass also applies to each of the component species, since there is no chemical reaction: Clearly the rate of change of all the salt in the tank equals the rate at which salt enters less the rate at which salt leaves. Thus, the balance equation for salt is

$$\frac{dc \, Ah}{dt} = c_f q_f - c q_e. \tag{4.2}$$

We could also write a balance equation for the water, but it would not be independent.

We may assume that the flow rates q_f and q_e and the density and concentration of the incoming stream are known, and of course the area is a constant. Equations 4.1 and 4.2 thus involve three characterizing variables: h, ρ, and c. We immediately see one important difference between single-component and multicomponent systems: The liquid level in the tank in a single-component system does not change with time if $q_f = q_e$ (Section 2.5). Because we have two components, even if $q_f = q_e$ there is no reason to suppose that $\rho = \rho_f$. (We might, for example, have a tank initially full of pure water, with $\rho = 1{,}000$ kg/m³, and an inlet stream of 26 percent aqueous NaCl with $\rho_f = 1{,}194$ kg/m³.) In that case the derivative on the left-hand side of Equation 4.1 will not equal zero.

Any problem in analysis requires a precise question; without a question we do not know how to construct the appropriate model. Let us identify our objective to be to determine how the tank level $h(t)$ varies as the inflow composition changes. Equations 4.1 and 4.2 are available to us, but these two equations involve three unknown variables, $\rho(t)$, $c(t)$, and $h(t)$. Referring to Figure 4.2, we must answer "No" to the query "Enough equations?" and "Yes" to the query "Conservation equations fully exploited?" (Conservation of energy and momentum are clearly irrelevant here, and there appears to be no more information to be obtained from mass conservation.) Thus, we are in need of a constitutive relation. There are no surprises here: We know that the density of a salt-water solution, or of any other

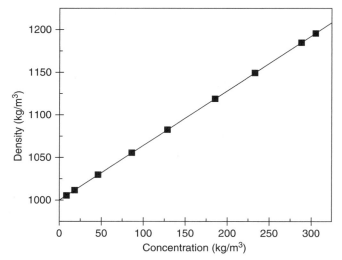

Figure 4.3. Density of aqueous NaCl solutions. The line was drawn using the least-squares method described in Appendix 2E.

multicomponent system, is a unique function of composition at a given temperature and pressure. Thus, we may write

$$\rho = \rho(c), \tag{4.3}$$

where the specific functionality $\rho(c)$ will depend on the system being investigated (i.e., on its *constitution*). Together with values of h and c at some time, say $t = 0$, we now have enough equations to address the question.

4.3 Changing Density

Now let us obtain a solution to the following question: How does the tank level change as a result of the changing feed conditions? Equation 4.3 is too general, but we have data available on salt-water systems to obtain a specific functional form. Figure 4.3 shows data plotted from a table in the *Chemical Engineer's Handbook*. There is some slight curvature to the data, but good straight lines can be drawn through the data over large regions and there is little error in drawing a single straight line through all the data shown here. We then write

$$\rho(c) = \rho_0 + bc, \tag{4.4}$$

where ρ_0 is an empirical constant that need not be the density of pure water, especially if we are interested in fitting a straight line to data that are far removed from $c = 0$. (In fact, ρ_0 differs from the density of pure water by only 0.1 percent in the least-squares fit to these data.) We can now substitute Equation 4.4 into Equation 4.1 to obtain (noting that $\rho_f = \rho_0 + bc_f$)

$$\frac{d}{dt}[\rho_0 + bc]Ah = [\rho_0 + bc_f]q_f - [\rho_0 + bc]q_e, \tag{4.5a}$$

or, making use of the fact that $\rho_0 A$ is a constant and doing a bit of rearranging,

$$\rho_0 A \frac{dh}{dt} = \rho_0 [q_f - q] + b \underbrace{\left\{ [c_f q_f - c q_e] - \frac{dc\,Ah}{dt} \right\}}_{= 0 \text{ from Equation (4.2)}}. \qquad (4.5b)$$

The terms multiplied by b in Equation 4.5 sum to zero because of Equation 4.2. Dividing both sides of the equation by $\rho_0 A$ thus leads to

$$\frac{dh}{dt} = \frac{q_f}{A} - \frac{q_e}{A}. \qquad (4.6)$$

This equation is identical to Equation 2.5, and we reach the interesting but unexpected conclusion that *the tank height is determined only by the volumetric flow rates, and the tank height is constant if $q_f = q_e$*. This result is only valid, of course, if the density is linear in concentration over the concentration range of interest. While we have derived this result only for a binary nonreacting system, it is, in fact, straightforward to extend the result to any single-phase nonreacting system for which the density is a linear function of all components. In a thermodynamics course, such a system would be called an *ideal solution*. When mixing the components of an ideal solution together, the total volume equals the sum of the volumes of the individual components.

It is quite common in reading the technical literature to find the assumption that all densities in a liquid phase system are equal, with the explicit or implicit assumption that the densities of all liquids are close to 1,000 kg/m^3. The consequence of such an assumption is that Equation 4.6 or its equivalent follows immediately. In fact, the densities of common liquids typically range from about 850 to 1,200 kg/m^3, and it is certainly incorrect to assume that numbers that differ by 40 percent are always "equal." The reason this "equal density" assumption works, leading to Equation 4.6, is for the reason shown here. This gross simplification is rigorous only for a linear density function. In general, we may write

$$\rho(c) = \rho_0 + bc + \varphi(c), \qquad (4.7)$$

where $\varphi(c)$ contains the nonlinear part of the relation; that is, the deviation from linearity. The resulting equation for h is then

$$\frac{dh}{dt} = \psi(c, c_f) \frac{q_f}{A} - \frac{q_e}{A}, \qquad (4.8a)$$

where

$$\psi(c, c_f) = \frac{\rho_0 + \varphi(c_f) - c_f \varphi'(c)}{\rho_0 + \varphi(c) - c\varphi'(c)}. \qquad (4.8b)$$

$\varphi'(c)$ is the derivative $d\varphi/dc$. In most cases, $\psi(c, c_f)$ will not differ from unity in a liquid system by enough to have any noticeable effect on the results relative to use of Equation 4.6.

4.4 Perfect Mixing Assumption

The assumption that the system is sufficiently well stirred to permit us to take the mixing to be perfect and instantaneous is not intuitively obvious and certainly needs validation. The simplest validation experiment that we can visualize is one in which a tank is initially filled with a salt-water solution at a concentration c_0. At time $t = 0$ we begin an inflow of pure water ($c_f = 0$) at a volumetric flow rate q and commence to pump the salt solution from the tank at the same rate ($q_f = q_e = q$). Under these conditions, it follows from Equation 4.6 that $dh/dt = 0$ and the volume $V = Ah$ occupied by the solution is a constant for the entire time of the experiment. Equation 4.2 then simplifies to

$$\frac{dc}{dt} = -\frac{q}{V}c \equiv -\frac{1}{\theta}c, \qquad (4.9)$$

where we have made use of the fact that $V = Ah$ is a constant and can be removed from the derivative. V/q has dimensions of time, and we denote the ratio by θ; θ is known as the *residence time*, and can be shown from probabilistic considerations to be the average time that a particle of mass spends in the tank if all particles have an equal probability of leaving at each instant of time (i.e., perfect mixing, where all particles have an equal probability of being at the exit at any time). Except for nomenclature ($h \rightarrow c$, $A/k \rightarrow \theta$) this is the same as Equation 2.12, so the solution is

$$\ln \frac{c(t)}{c_0} = -t/\theta. \qquad (4.10)$$

Figure 4.4 shows data of salt concentration (measured by electrical conductivity) as a function of time for an experiment designed to test this analysis, but with a mixer intentionally selected to be inefficient. The data at 1,000 RPM follow the theoretical line reasonably well, showing that with sufficient agitation the predictions based on perfect mixing agree with experiment. The data at 100 RPM show poor agreement; the poor agreement cannot be interpreted without more information about the precise placement of the conductivity probe, but it illustrates what can happen with inadequate agitation.

Many mixing experiments have been carried out in a variety of geometries with different fluids and impellers, and design procedures are available for selecting the impeller geometry and motor size for most situations with low-viscosity (waterlike) liquids. Mixing remains an active research area, with the focus on mixing multiphase systems (gas/liquid, solid/liquid) and high-viscosity liquids. The mixing of granular materials (fine catalysts or powders, for example, or even beach sand), which are liquidlike in behavior when flowing, is of particular importance in the pharmaceutical industry, where the precise distribution of components is essential to ensure proper dosages in every pill or capsule.

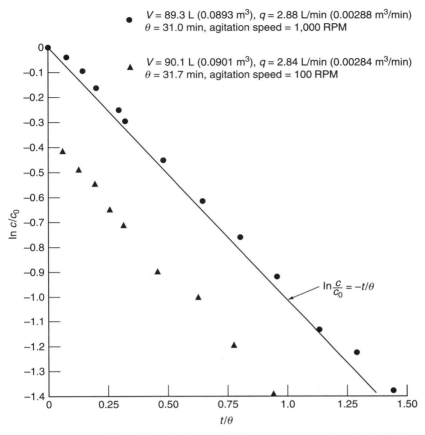

Figure 4.4. Natural logarithm of c/c_0 vs t/θ for an aqueous salt solution being washed from a 67-cm-diameter tank with a liquid height of 25.4 cm and equal inflow and effluent rates. Agitation was by a 6.35 cm-diameter marine propeller. • $V = 89.3$ L (0.0893 m³), $q = 2.88$ L/min (0.00288 m³/min), $\theta = 31.0$ min, agitation speed = 1,000 RPM. ▲ $V = 90.1$ L, $q = 2.84$ L/min, $\theta = 31.7$ min, agitation speed = 100 RPM. Data collected by D. T. T. Chang.

4.5 Air Quality and Rule-of-Thumb Design

4.5.1 Introduction

Exposure to toxic substances is a major concern in both the workplace and the home. Some substances, such as carbon monoxide (CO), are routinely monitored in laboratories to ensure that dangerous levels are not reached. Indeed, many chemical engineers are involved in the design of sensors, and the development of new methodologies utilizing microprocessing technology with component-specific surface sites is an active area of chemical engineering research.

Exposure to most toxic substances occurs in environments where the substances are nonreactive; the problems occur only when they are taken into the body. CO does not react in air at normal room temperature and pressure, for example; it is only when it enters the bloodstream and complexes with hemoglobin to form carboxy-hemoglobin, reducing the oxygen-carrying capacity of the blood, that its reactivity becomes relevant. Because of this nonreactivity, we can analyze the transport and buildup of CO and many other toxic substances using the methodology developed in this chapter, illustrating a simple but very important application of component mass balances. Air is compressible, but we can assume that the small changes in density resulting from the very small pressure variations required to pump air using fans or blowers can be neglected in our analysis.

4.5.2 Laboratory Ventilation

Laboratory ventilation is an important factor in a chemistry or chemical engineering environment. The U.S. Occupational Safety and Health Administration (OSHA) Standard 1910.1450, Appendix A, recommends four to twelve room air changes per hour in a laboratory if fume hoods are used as the primary means of controlling toxic materials. This means that if the volume of the laboratory is V m^3, the flow rate of fresh air to the laboratory must be at least $4V$ m^3/hr. A laboratory with floor space 10 m × 10 m and a ceiling height of 4 m, or a volume of 400 m^3, would then require a ventilation rate of at least 1600 m^3/hr, or 0.44 m^3/s.

Suppose a volatile toxic substance is put away improperly and escapes into the room at a constant rate of S kg/s. If we assume that dispersion in the room is sufficiently rapid that we can assume perfect mixing (a questionable assumption, but useful for purposes of discussion) and that the inflow air contains none of the toxic substance, the component mass balance is then

$$V\frac{dc}{dt} = S - qc, \tag{4.11}$$

where c is the concentration of the toxic material. This equation is easily solved, most simply by changing the independent variable to $u = c - S/q$, in which case Equation 4.11 becomes $du/dt = -u/\theta$, where $\theta = V/q$. The solution is then $u(t) = u_0 \exp(-t/\theta)$, where $u_0 = -S/q$ if there is none of the toxic substance in the room air at time $t = 0$. The solution can then be written

$$c(t) = \frac{S}{q}[1 - e^{-t/\theta}]. \tag{4.12}$$

According to the OSHA guideline for minimum flow, $\theta = V/q = V/4V = 0.25$ hr. $e^{-t/\theta}$ is essentially zero when t/θ is greater than about 4; at that time the system will have reached a steady state $(dc/dt = 0)$, with $c = S/q$. Under the minimum OSHA guideline the steady state will be attained in about one hour. If the volatile compound evaporates at a rate $S = 4$ g/hr and $q = 1,600$ m^3/hr, then the steady-state concentration in the air will be 2.5×10^{-6} kg/m^3. The density of air is about 1.2 kg/m^3, so the volatile toxic material will reach a level of about 2 ppm $(2 \times 10^{-6}$ kg/kg of air). Prolonged exposure to some toxic materials at this level could be hazardous. It is important to note that many hazardous materials are denser than air and may have concentrations at floor level that are greater than what would be estimated using the perfect mixing assumption.

4.5.3 Airflow Design

Anyone who has traveled in a commercial airplane has experienced a "stuffy" feeling from the reduced oxygen level and the accumulation of CO_2 resulting from too little fresh airflow. The energy cost of bringing fresh air into a commercial jetliner at a 10,000-meter elevation is substantial, and the airflow is a balance between cost and passenger comfort. Similar considerations apply to room ventilation in a building. Laboratories and modern offices typically have forced-air systems that ensure a specified level of air replacement, as noted in the preceding section. Most homes, on the other hand, are designed to minimize air replacement, and the only fresh air that enters is the result of "leaks" in a system that the homeowner wishes to be airtight.

Let us consider a room of volume V m^3 occupied by N people. The flow of fresh air to the room is q m^3/min. Each individual is adding a nonreactive species A to the room air at a rate w_A mg/hr. If A is CO_2, for example, the metabolic rate for the average person with daily light activity is 200 cm^3/min at standard temperature and pressure (STP), which is equivalent to 21,600 mg/hr of CO_2. We wish to determine the airflow necessary to keep the concentration c_A below the level determined to be safe.

Our control volume is the room, *excluding the space occupied by the people.* We need to keep the people outside the control volume for two reasons. First, they are not part of the volume occupied by the room air, so including them could introduce an error into the calculation of concentrations. This is a minor factor, however. The more important reason is that the substances of interest are nonreactive in the room but not in the people. By excluding the people from the control volume, w_A is simply a mass flow rate into the control volume, and we need not be concerned with the chemistry of metabolism.

The unoccupied volume of the room will be presumed to be unchanged, and the temperature and pressure remain constant, so the mass flow rates in and out must be equal. We can assume that the net difference in volumetric flow of gases into and out of the people is negligible,[*] so the volumetric airflow in and out must be the same. We will also assume that V refers to the unoccupied volume of the room, but this will not differ significantly from the total room volume.

We assume that air circulation in the room is sufficient to make the perfect mixing assumption. The mass balance for species A then simply states that the rate of change of the mass of A in the room equals the rate at which A enters (Nw_A from

---

[*] This is a subtle point. For CO_2 production, every mole of O_2 entering the body is balanced by a mole of CO_2 leaving, and at STP the two gases have the same volume. If we are considering CO production by smokers, however, each mole of O_2 leaving is replaced by two moles of CO entering, hence twice the volumetric flow rate. We presume that these numbers are so small that they can be ignored relative to q.

the N people and qc_{Af} from the inflow air stream) less the rate at which A leaves (qc_A):

$$\frac{dVc_A}{dt} = V\frac{dc_A}{dt} = qc_{Af} + Nw_A - qc_A. \qquad (4.13)$$

We first consider the steady-state design problem, where we take $dc_A/dt = 0$. We can the solve for the flow rate, q:

$$q = \frac{Nw_A}{c_A - c_{Af}}. \qquad (4.14)$$

This may be, on first glance, a surprising result, for it tells us that the airflow rate is independent of the size of the room but is proportional to the number of people. Thus, the design variable is q/N m^3/hr/person, and we need to know the typical occupancy level of the room.

For CO_2 we typically have an individual respiration rate $w_A = 21{,}600$ mg/hr. Let us suppose that we are willing to accept an increase in CO_2 level of 400 mg/m^3 over the level in the incoming air. (400 mg/m^3 corresponds to about 200 ppm. The total level of gases other than nitrogen, oxygen, and argon in air is typically around 700 ppm.) We then have $q/N = 21{,}600/400 = 54$ m^3/hr/person. The 1989 American Society of Heating, Refrigerating, and Air-Conditioning Engineers (ASHRAE) standard for industrial offices is 20 ft^3/min/person, or 34 m^3/hr/person, which is somewhat less than our calculation and corresponds to a net increase of 635 mg/m^3.

The production of CO from cigarette smoke is a serious environmental problem that has led to legislation banning cigarette smoking in many public places. The typical cigarette produces 25 to 50 mg of CO and 160 to 480 mg of CO_2. Let us consider CO, which we will denote as species B, and we will use the higher number of CO production to ensure conservative estimates. We will assume that the typical smoker smokes three cigarettes per hour, so $w_B = 3 \times 50 = 150$ mg/hr. We can assume there is no CO in the fresh air ($c_{Bf} = 0$), so the design equation for airflow is $q/N = 150/c_B$. The ASHRAE standard of $q/N = 34$ m^3/hr for office ventilation will result in a CO concentration of $c_B = 150/34 = 4.4$ mg/m^3. This is about a factor of ten below the U.S. Environmental Protection Agency (EPA) warning level of 40 mg/m^3, where prolonged exposure can lead to "changes in myocardial metabolism and possible impairment; statistically significant diminution of visual perception, manual dexterity, or ability to learn," so it is probably a safe level. The CO_2 level would also increase because of the smoke, and of course we have not considered other components of the smoke, such as tars, or the discomfort to other occupants.

In a specially designed smoking lounge we might expect a much higher smoking rate, in part because any one individual will only spend short periods in the room. Let us suppose that the typical smoking rate in a lounge is six cigarettes per hour, in which case $w_B = 6 \times 50 = 300$ mg/hr. If we suppose that we wish to keep the CO level to a maximum of 4 mg/m^3 (10 percent of the EPA limit) we have $q/N = 300/4 = 75$ m^3/hr/person. The ASHRAE standard for smoking lounges is 50 ft^3/min/person, or 85 m^3/hr/person, so the standard assumes either a slightly lower CO level in the room (3.5 mg/m^3) or a slightly higher smoking rate.

4.5.4 Rules of Thumb in Design

The design problem we have been dealing with here is a very elementary one, but it does illustrate a general point. Designs are often carried out using rules of thumb, which are general principles that summarize extensive experience. These rules of thumb often have a rational basis and can be derived by relatively straightforward analysis of the situation at hand. The ASHRAE standards for office and smoking lounge airflow rates are rules of thumb, and they are routinely applied by engineers and architects, but it is clear they correspond to nothing more than some basic data, information about acceptable levels of specific toxic substances and, in the case of CO, an assumption about the habits of the typical smoker. The practicing chemical engineer is usually faced with far more complex design decisions and more opaque rules of thumb, but the same principle – that collected experience is often amenable to analysis – still applies.

4.5.5 Transient Indoor Air Quality

We have focused thus far on the airflow design problem, which is based on steady-state levels. The transient calculation is also of interest, especially with regard to home air quality. We will assume in Equation 4.13 that c_{Af}, q, and Nw_A are constants; in that case we have seen this form of equation many times, and we can immediately write the solution with $c_A = c_{A0}$ at $t = 0$ as

$$c_A = c_{A0}e^{-t/\theta} + \left[c_{Af} + \frac{Nw_A}{q}\right]\left[1 - e^{-t/\theta}\right], \qquad (4.15)$$

where $\theta = V/q$.

Suppose the typical office setting provides 18 m³/person (3 m × 2 m floor space and a height of 3 m). The ASHRAE standard of 34 m³/hr/person then gives $\theta = 18N/34N = 0.53$ hr = 32 min. Thus, the mean residence time is such that the air is replaced about two times every hour. If we use the same space allocation in a smoking lounge, with $q = 85N$, we find $\theta = 0.2$ hr = 12 minutes, and the air is replaced five times per hour, comparable to the lower end of the standard for a laboratory.

Now let us consider three smokers in a closed room in a home. A typical room size might be 3 m × 5 m × 3 m, or 45 m³. If we take the extreme case of no air flow ($q = 0$) it follows from Equation 4.13 (replacing A with B to denote CO) that

$$q = 0 : c_B = c_{B0} + \frac{Nw_B}{V}t. \qquad (4.16)$$

(The same result follows from Equation 4.13 by taking the limit $q \to 0$, recalling that $\theta = V/q$ and applying L'Hôpital's rule to the indeterminate term on the right.) If we suppose that they are each smoking three cigarettes per hour and $c_{B0} = 0$, we have $c_B = [(3 \times 150)/45]t = 10t$, where t is in hours and c_B is in mg/m³. Thus, after one hour the CO level would reach 10 mg/m³, and if they were to continue for a total of four hours the CO would reach a level at which prolonged exposure is dangerous.

We now suppose that the house is extremely well insulated in order to restrict airflow, but that some ventilation does occur. q might be as low as 5 m^3/hr, so $\theta = V/q = 45/5 = 9$ hours. In Equation 4.15, with $c_{B0} = c_{Bf} = 0$ and $Nw_B/q = 90$ mg/m^3, we find $c_B = 9.4$ mg/m^3 after one hour and 31 mg/m^3 after 4 hours; the short-time result ($t/\theta \ll 1$) at $t = 1$ hour is close to the result for $q = 0$, with poorer agreement at longer time. This is because Equation 4.16 follows directly from Equation 4.15 if we expand the exponential in a series for small values of t/θ and truncate after the linear term.

4.6 Concluding Remarks

This short chapter contains a number of very important concepts. Perfect mixing is frequently assumed when analyzing flowing systems; as we see here, the validity of the assumption depends on the physical properties of the fluids and the design of the agitation system, but it can often be achieved. Linearity of the density–concentration constitutive equation (an ideal solution) is often assumed implicitly, sometimes with the unnecessary statement that all densities are assumed to be equal. The notion of a rule of thumb is extremely important. Many practicing engineers use rules of thumb for estimation and design. What we see here in our simple ventilation example is that rules of thumb can often be derived from first principles analyses; the ASHRAE standards quoted here are comparable to the results of our analysis, with small differences in the assumptions about individual behavior. The material that follows in subsequent chapters will utilize the ideas developed here and in the preceding chapters in a series of engineering applications.

Bibliographical Notes

Mixing technology – the selection of mixers for various applications, energy requirements, and so forth – is covered in standard chemical and mechanical engineering handbooks. The *science* of mixing, which deals with the details of the temporal and spatial distribution of the flow, is a specialized topic in fluid mechanics that would normally be studied in an advanced course, probably at the graduate level. Broad treatments of mixing intended for practitioners are in

Harnby, N., M. F. Edwards, and A. W. Nienow, Eds., *Mixing in the Process Industries*, 2nd Ed., Butterworth-Heinemann, London, 1997.
Paul, E. L., V. Atiemo, and S. M. Kresta, Eds., *Handbook of Industrial Mixing: Science and Practice*, Wiley-Interscience, New York, 2003.

There are many books on indoor air quality and ventilation systems. One whose author has a first degree in chemical engineering is

Bearg, D. W., *Indoor Air Quality and HVAC Systems*, CRC Press, Boca Raton, FL, 1993.

Design rules of thumb are a main focus of

Bell, A., Jr., *HVAC Equations, Data, and Rules of Thumb*, 2nd Ed., McGraw-Hill, New York, 2007.

A collection of rules of thumb for chemical engineers, most of which address issues that appear later in the curriculum or in professional practice, is

Brannan, C. R., *Rules of Thumb for Chemical Engineers*, 4th Ed., Gulf Publishing, Houston, 2005.

PROBLEMS

4.1. A well-stirred process vessel contains 1.7 m^3 (450 gallons) of a solution of water and 34 kg of sodium chloride at the time that a pure water feedstream is introduced. Find the salt concentration at the end of sixty minutes if the water flow to the tank is maintained at 0.0265 m^3/min and brine is removed at the same rate.

4.2. The Mediterranean Sea exchanges water with the Atlantic Ocean. Fresh water inflow to the Mediterranean is approximately 30,000 m^3/s and evaporation occurs at a rate of approximately 80,000 m^3/s. The salt content of the Mediterranean is 37 g salt/1,000 g solution and it is 36 g/1,000 g in the Atlantic. Estimate the flow from sea to ocean and from ocean to sea.

4.3. According to OSHA, a level of chlorine in the air of 3 ppm will cause impairment. Suppose there is a constant source of chlorine of 4 g/hr in a room that is ventilated at a rate of 250 m^3/s. What is the steady-state concentration of chlorine in the room? Is it above or below the level that will cause impairment?

4.4. The OSHA recommendation for laboratory ventilation is four to twelve room air changes per hour. Approximately how long does it take to reach a steady-state concentration in a laboratory with a constant contaminant source if the air is replaced eight times per hour? How long does it take to get 90 percent of the way to steady state?

4.5. Four people enter an empty smoking lounge that has a volume of 50 m^3 and is ventilated at a rate of 120 m^3/hr. Each person smokes about two cigarettes an hour, and each cigarette produces about 40 mg of CO. Assume that there was no CO in the lounge when they entered.

    a.   What is the concentration of CO in the lounge at the end of 30 minutes? 1 hour?

    b.   How long will it take for the CO concentration to reach a steady state? What will the steady-state concentration be?

4.6. A soft drink process requires mixing sugar and water continuously in a 250 m^3 tank. We have one measurement of the density of sugar: At a concentration of 207.5 kg/m^3 the density is 1,083 kg/m^3. The process operators measure the concentration of sugar by weighing a fixed volume of solution, and they have a correlation chart that is a straight line passing through zero at a density of $1,000 \ kg/m^3$.

    a.   The plant is being shut down for maintenance, so the tank is flushed with pure water at a rate $q = 1 \ m^3/s$. The initial concentration of sugar is $100 \ kg/m^3$. What are the density and concentration in the tank after 5 minutes, assuming that the operators' chart is correct?

    b.   How long should the operators expect to have to wait for the sugar concentration to be below 1 ppm?

4.7. A well-dispersed mixture of solids and liquid is to be concentrated by being pumped through a tank with a porous plate at the bottom through which pure liquid may pass, as shown schematically in Figure 4P.1. Because the dispersion is very good you may consider the mixture to be homogeneous liquid with a solids concentration c.

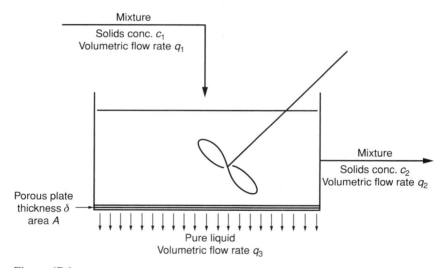

Figure 4P.1.

    a.   Show that the mixture density is a linear function of c. The total volume V is the sum of the solids volume V_S and the liquid volume V_L.

    b.   It is observed experimentally that the logarithm of height varies linearly with time when pure water flows through the porous plate. Assume that the volumetric flow rate q_3 depends on the liquid density and viscosity, ρ_L and η_L, respectively; the plate area, A; and the pressure change per unit thickness across the plate, $\Delta p/\delta$. Obtain a constitutive equation for q_3.

    c.   Obtain a steady-state design equation relating the tank dimensions to the liquid throughput and the desired degree of concentration.

5 Membrane Separation

5.1 Introduction

Most processes, both physicochemical and biological, involve one or more separation process. Human physiology, for example, requires the transfer of oxygen from air in the lungs to the blood stream, and the simultaneous transfer of CO_2 from the blood stream to the lungs for removal. The function of the kidney is to process a continuous flow of liquid in order to separate waste products for removal from the body. The production of ethanol by the fermentation of sugars produced from natural products, whether for energy applications or for whiskey, requires that the ethanol be separated from an aqueous stream in which ethanol comprises less than 20 percent. The manufacture of polyethylene terephthalate for textile fibers requires the removal of ethylene glycol that is produced during the condensation polymerization process. A DNA analysis requires the separation of DNA fragments with different lengths and base pairs by gel electrophoresis. The production of oxygen for industrial or medical applications requires that the oxygen be separated from an air stream.

There are a variety of separation methodologies, and the analysis of separation processes has historically held a prominent place in the chemical engineering curriculum. Some, such as distillation and extraction, are familiar. (Brewing coffee or tea is an extraction process that is carried out on a very small scale.) Most physiological separations are membrane processes, in which a thin membrane keeps fluid (liquid or gas) streams apart while permitting certain species to move across the membrane. Membrane separations are playing an increasingly important role in technological applications, including water purification and gas separation, and this chapter is intended to serve as a brief introduction while amplifying our understanding of the application of mass balances.

Most synthetic membranes are polymeric, although ceramic, zeolite, and metal membranes are used in some applications. The mechanism for the preferential transport of a particular species across a membrane depends on a variety of factors, including the size of the pores (if any), the detailed morphology of the membrane, and the energy of interaction between the species and the membrane material.

All other things being equal, the rate of transport is proportional to the available surface area, as would be expected. The rate also depends on a *driving force* that reflects the difference in the thermodynamic states of the species in the two fluids on the opposite sides of the membrane. The notion of thermodynamic state requires concepts that are typically studied later in the curriculum, but we can address one important case. Suppose we have dilute aqueous streams on both sides of the membrane, and we wish to transport a dissolved species across the membrane. In the absence of any reaction (chemical complexation, for example) the transport will proceed until the concentration of the species is the same on both sides of the membrane. To a good approximation, the rate at which the system tries to equilibrate the concentrations is proportional to the *difference* between the concentrations on the two sides of the membrane. The proportionality constant per unit area of membrane surface is known as the *permeability*,[*] which we denote Π; Π will have dimensions of length/time in our formulation. (The dimensions of permeability are determined by the driving force, which is sometimes partial pressure; hence, permeability can sometimes have unusual dimensions. For gas separation membranes the usual units are $mol/m \cdot s \cdot Pa$.)

5.2 Single-Stage Dialysis

Membrane separation of one or more dissolved species from a dilute aqueous stream is known as *dialysis*. The "artificial kidney," which substitutes periodically (typically three times per week) for the kidney function in patients with end-stage renal disease, is a dialysis process in which 250–400 cm^3/min of blood is passed through the membrane system for the removal of urea and other toxins. Dialysis is also used in emergency situations to remove drugs or poisons from the system. Most dialysis processes consist of long flat or tubular membranes in channels with large aspect ratios, but it is convenient for our purposes to start with an idealized situation in which two well-mixed volumes are separated by a membrane, and a dissolved species is transported across the membrane because of a concentration difference, as shown schematically in Figure 5.1. Water may also migrate across the membrane because of the different states on the two sides, but we will ignore this phenomenon because the amount of water transfer will generally be small in the absence of a significant pressure difference between the two sides of the membrane.

The stream from which we wish to remove the dissolved species is known as the *raffinate* (from the French *raffinere*, to refine); we denote the volumetric flow rate of the raffinate by R and the concentration of the dissolved species by c_R. The stream to which the species is transferred is known as the *permeate* (a noun form

---

[*] This description is a bit simplistic, but it suffices for our introductory discussion here. In reality, the resistance consists of the membrane permeability *per se,* and resistance to mass transfer *within each fluid phase* in the neighborhood of the surface; the latter resistance depends on the flow rates and certain physical properties of the particular species. The actual membrane permeability may contribute less than half of the total resistance in liquid systems. Nonetheless, the functional form will remain the same; it is only the calculation of Π that is affected.

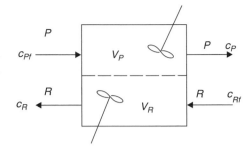

Figure 5.1. Schematic of a single-stage dialysis process, with two well-stirred sections separated by a semipermeable membrane.

not found in standard dictionaries of English usage, from the Latin verb *permeare,* to go through); we denote the volumetric flow rate of the permeate by P and the concentration by c_P. We assume that the amount of material transferred across the membrane is too small to affect the volumetric flow rates, so the outflow values of P and R, respectively, are the same as the inflow values. (This assumption is not necessary, but it is realistic and it simplifies the algebra greatly, making it easier to understand the key features of the process.) The volumes are denoted V_R and V_P, respectively, and the area of the membrane is denoted A. The rate at which the dissolved species leaves the raffinate and enters the permeate is then $\Pi A(c_R - c_P)$.

Each volume, V_R and V_P, is a distinct control volume, and we write the balance equations separately for each control volume. The overall mass balances are replaced by the assumptions that the raffinate and permeate flow rates into and out of the stages are the same. The balance equations for the dissolved species in the two volumes are then as follows:

$$V_R\frac{dc_R}{dt} = R(c_{Rf} - c_R) - \Pi A(c_R - c_P), \tag{5.1a}$$

$$V_P\frac{dc_P}{dt} = P(c_{Pf} - c_P) + \Pi A(c_R - c_P). \tag{5.1b}$$

The first term on the right side of each equation represents the difference between the inflow and outflow rates of the dissolved species, whereas the second term represents the rate of transfer across the membrane. The transfer term is negative in the raffinate equation and positive in the permeate equation, since solute transfer is from the raffinate (outflow) to the permeate (inflow). We are interested in the design equations, for which it suffices to consider only the steady state, so we set $dc_R/dt = dc_P/dt = 0$ and get

$$R(c_{Rf} - c_R) - \Pi A(c_R - c_P) = 0, \tag{5.2a}$$

$$P(c_{Pf} - c_P) + \Pi A(c_R - c_P) = 0. \tag{5.2b}$$

Equations 5.2a and 5.2b can be added together to obtain a simple relation between the concentrations, which can be rearranged to the form

$$c_P = c_{Pf} + \frac{R}{P}(c_{Rf} - c_R). \tag{5.3}$$

We can then solve Equation 5.2a to obtain the net separation in the raffinate stream,

$$c_{Rf} - c_R = \left[\frac{1}{1 + \frac{R}{P} + \frac{R}{\Pi A}} \right] (c_{Rf} - c_{Pf}) \equiv M(c_{Rf} - c_{Pf}). \tag{5.4}$$

Equation 5.4 is the basic design equation for the single-stage dialyzer. Notice that some useful estimates follow immediately. First, consider the case $A \to \infty$; that is, there is no limit on the available membrane surface area for transport. Equation 5.4 can then be solved to obtain the minimum value of P/R that can ever be used (i.e., the lower bound, which is never attainable in practice) in order to effect a given separation:

$$\left(\frac{P}{R} \right)_{min} = \frac{c_{Rf} - c_R}{c_R - c_{Pf}}. \tag{5.5}$$

From Equation 5.3, we see that this limit corresponds to $c_P = c_R$; that is, to equal concentrations on both sides of the membrane, which is clearly the best that we can do. If the flow rate R of the raffinate stream is specified, as will often be the case, then Equation 5.5 leads to an explicit expression for the lower bound on the minimum permeate rate:

$$P_{min} = R \frac{c_{Rf} - c_R}{c_R - c_{Pf}}. \tag{5.6}$$

Similarly, the minimum surface area will correspond to the case in which the permeate flow is so rapid that the transferred material is removed instantaneously ($P/R \to \infty$), causing the driving force to be maximized; we therefore obtain

$$A_{min} = \frac{R}{\Pi} \left[\frac{c_{Rf} - c_R}{c_R - c_{Pf}} \right] = \frac{P_{min}}{\Pi}. \tag{5.7}$$

We can usually assume that $c_{Pf} = 0$, so these limits can be expressed solely in terms of the *separation factor*, $s_R = c_{Rf}/c_R$: $P_{min} = R(s_R - 1)$ and $A_{min} = R(s_R - 1)/\Pi$.

5.3 An Optimal Design Problem

Engineering design frequently requires the solution of an optimization problem, in which a profit is maximized or a cost is minimized. The single-stage dialysis system is not an efficient engineering separation system, as we shall see in subsequent sections, but it does provide a good framework for looking at how an optimal design might be carried out. The same approach would be employed for the more efficient configurations that we will develop in later sections.

The total cost of the system includes the capital cost for construction and the net present value of the operating cost. Capital cost is related to size, and for illustrative purposes here we will take the capital cost to be proportional to the total area. Operating costs are related to throughput. If R and the required separation are specified, as will often be the case, then the variable operating costs are determined by the value of P, and we will again assume that the costs are proportional to P. We therefore have a cost function of the form

$$C = A + \Lambda P, \tag{5.8}$$

where Λ reflects the relative weight of capital and operating costs. Note that Λ has dimensions of time/length. There will be other terms that represent fixed costs, but because they are fixed they do not enter into the optimization calculation. The linear relationship is very unrealistic, but it suffices for illustrative purposes and permits us to carry through the steps for a solution without unnecessary computational complexity. The design problem is now to find A and P such that we minimize C.

The simplest way to solve this minimization problem is first to obtain A in terms of P. We assume $c_{pf} = 0$, in which case we can rewrite Equation 5.7 in terms of the separation factor $s_R = c_{Rf}/c_R$ as

$$A = \frac{R}{\Pi} \frac{s_R - 1}{1 - \frac{R}{P}(s_R - 1)}. \tag{5.9}$$

(Note that $A > 0 \Rightarrow P > P_{min} = R(s_R - 1)$, as required.) We then write C as a unique function of a single variable, the unknown permeate flow rate P:

$$C = \frac{R}{\Pi} \frac{s_R - 1}{1 - \frac{R}{P}(s_R - 1)} + \Lambda P. \tag{5.10}$$

C becomes infinite as P goes to either of the two limiting values, P_{min} or infinity, so C must have a finite minimum for some intermediate value of P. The minimum is obtained by setting the derivative dC/dP to zero*, as follows:

$$\frac{dC}{dP} = \frac{R^2}{\Pi P^2}(s_R - 1)^2 \frac{1}{\left[1 - \frac{R}{P}(s_R - 1)\right]^2} + \Lambda = 0, \tag{5.11}$$

which, when solved for the optimal P, gives

$$P = \frac{R(s_R - 1)\left(1 + \sqrt{\Lambda \Pi}\right)}{\sqrt{\Lambda \Pi}} = P_{min} \frac{1 + \sqrt{\Lambda \Pi}}{\sqrt{\Lambda \Pi}}. \tag{5.12a}$$

The corresponding area is then

$$A = \frac{R(s_R - 1)\left(1 + \sqrt{\Lambda \Pi}\right)}{\Pi} = A_{min}\left(1 + \sqrt{\Lambda \Pi}\right). \tag{5.12b}$$

As $\sqrt{\Lambda \Pi} \to 0$, $A \to A_{min}$ and $P \to \infty$, as expected, while $P \to P_{min}$ and $A \to \infty$ as $\sqrt{\Lambda \Pi} \to \infty$. We would follow the same procedure for more realistic cost functions, but the calculus would be more complex and we might not be able to obtain analytical solutions. The absence of an analytical solution is not a problem in practice, since the minimum can be found using numerical methods, but the availability of an analytical solution makes it easier to understand issues such as sensitivity to parameters.

---

* Normally we must also check the sign of the second derivative to distinguish between a minimum, a maximum, or a saddle, but that step is not necessary here because we know that C has a minimum.

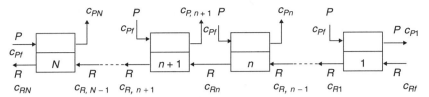

Figure 5.2. "Cross-flow" multistage dialysis.

The fact that the optimal solution for $A/A_{\min}$ is a function of $\sqrt{\Lambda\Pi}$ is to be expected from dimensional considerations.

5.4 Multistage Dialysis

The fact that the minimum membrane area (hence the capital cost) for single-stage dialysis depends only on the separation factor suggests a conceptual design for carrying out a separation more efficiently. The area for the single-stage dialyzer will increase by a factor of ten if we increase the desired separation factor by a factor of ten (e.g., we go from 90% separation, with $s_R = = 10$, to 99% separation, with $s_R = 100$). We could achieve the same result, however, by using two stages, each with the same separation factor, using the raffinate leaving the first stage as the feed to the second. With a separation factor of ten in each stage, for example, we would achieve 90 percent removal in the first stage and a further 90 percent removal in the second, giving us 99 percent removal overall, but with only twice the area of a single stage; the saving is somewhat offset by the doubling of the permeate flow, since each stage, in this concept, will require a separate permeate stream.

Let us consider a more general case, as shown in Figure 5.2, in which we employ N stages. If we focus on a typical stage, denoted by n, the raffinate feed to the stage is $c_{R,n-1}$, whereas the raffinate effluent is $c_{R,n}$. For simplicity we will assume that the permeate feed to each stage is pure, so $c_{Pf} = 0$. We can then immediately write down the separation in stage n by changing the nomenclature in Equation 5.4, rearranging slightly to obtain

$$c_{R,n} = (1 - M)c_{R,n-1}. \tag{5.13}$$

Equation 5.13 is a *finite difference equation*. Finite difference equations arise frequently in staged separations applications. The solution to this equation is easily seen by inspection: $c_{R0} = c_{Rf}$, so $c_{R1} = (1-M)c_{Rf}$, $c_{R2} = (1-M)c_{R1} = (1-M)^2 c_{Rf}$, and so on. The general result is clearly

$$\frac{c_{R,n}}{c_{Rf}} = (1 - M)^n = \left(\frac{1 + \frac{\Pi A}{P}}{1 + \frac{\Pi A}{P} + \frac{\Pi A}{R}} \right)^n. \tag{5.14}$$

This is not an easy equation to interpret. We can get some useful insight into the behavior by considering the case in which we have a large number of stages ($N \gg 1$), each of which is small, with a total membrane area $A_T = NA$. The permeate flow

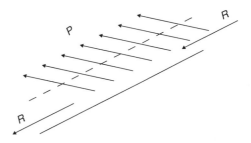

Figure 5.3. Schematic of a continuous cross-flow dialyzer, with uniform permeate flow transverse to the channel carrying the raffinate.

across each small stage will then also be small, and we define the total permeate flow as $P_T = NP$. We can then write the separation factor for the entire system as

$$s_R = \frac{c_{Rf}}{c_{R,N}} = \left[1 + \frac{\Pi A_T}{R \left(1 + \frac{\Pi A_T}{P_T} \right) N} \right]^N. \tag{5.15}$$

Now, $\lim_{N \to \infty} (1 + \frac{x}{N})^N = e^x$. Thus, for large N we can get a very good approximation to the separation factor by taking the limit $N \to \infty$ to obtain

$$s_R = \exp \left[\frac{\Pi A_T}{R \left(1 + \frac{\Pi A_T}{P_T} \right)} \right]. \tag{5.16}$$

This equation can be rearranged to solve for the total area, giving

$$A_T = \frac{R/\Pi}{\frac{1}{\ln s_R} - \frac{R}{P_T}}, \tag{5.17}$$

and the minimum area is

$$A_{T,\min} = \frac{R}{\Pi} \ln s_R = \frac{2.3 R}{\Pi} \log s_R. \tag{5.18}$$

Thus, each factor of ten in the separation factor requires an equal area; that is, a separation factor of 100 (99% removal) requires twice the area required for a separation factor of ten (90% removal), whereas a separation factor of 1,000 (99.9% removal) requires three times the area required for a separation factor of ten.

We should briefly consider how we might construct a dialyzer that is described by this equation. The notion that we have a large number of very small well-mixed regions suggests a flow channel that is long and thin, so there can be a great deal of local mixing but little longitudinal mixing over a significant length scale. (Such a device is said to exhibit *plug flow*.) The permeate would need to flow uniformly over the membrane in a transverse direction, as shown schematically in Figure 5.3. Such a device could easily be constructed; some hollow fiber devices, in which one stream flows through tubes with a membrane shell and the other stream flows across the tubes, approximate this configuration, and the design does not differ substantively from the way in which some physiological transport takes place in the body.

5.5 Optimal Permeate Distribution

The cross-flow separation analyzed in the preceding section suggests another nice, but simple, example of an optimization problem. We assumed that the flow rate of permeate to each stage was the same. This assumption is not intuitively obvious: Might we not do better by using more of the permeate in the first few stages, thus reducing the raffinate concentration more quickly, and then use less in the latter stages where there is less material to remove?

Let us suppose that the permeate flow rate to the n-th stage is P_n, and that the total flow rate is $\sum_{n=1}^{N} P_n = P_T$, but that we no longer assume that $P_n = P_T/N$ for all n. We will continue to assume that all stages have the same membrane area. It is straightforward to show from Equation 5.13 that

$$\frac{c_{R,n}}{c_{Rf}} = \prod_{k=1}^{n} \left(\frac{1 + \frac{\Pi A}{P_k}}{1 + \frac{\Pi A}{P_k} + \frac{\Pi A}{R}} \right), \tag{5.19}$$

where the symbol $\prod_{k=1}^{n}$ denotes *multiply all terms together, with k varying from 1 to n*. With a bit of manipulation we can then write the separation factor as

$$s_R = \frac{c_{Rf}}{c_{R,N}} = \prod_{k=1}^{N} \left[1 + \frac{\Pi A}{R \left(1 + \frac{\Pi A}{P_k} \right)} \right]^{-1}. \tag{5.20}$$

The optimization problem is now to choose the distribution of permeate flows $\{P_k\}$ in such a way that we maximize the separation factor, keeping in mind that the sum of all flows must equal P_T.

We show in Appendix 5.B how to solve a class of optimization problems that includes the one stated here. Before looking at the Appendix, however, we might try to reason out the solution. It is clear from Equation 5.20 that the order of multiplication is irrelevant; that is, every stage contributes in the same way, and, for a given distribution, we could renumber the stages without affecting the outcome. (That is, if we established a particular sequence from 1 to N, we could implement the same sequence from N to 1 without changing the separation factor.) This observation is counterintuitive, but it is clearly supported by the analysis. Hence, it should be true that the best distribution is one that is independent of the sequencing of the stages; that is, we expect that the optimal distribution will be equal flow rates to all stages. This is indeed the result that we obtain from the mathematical analysis in Appendix 5.B.

5.6 Smart Engineering: Countercurrent Dialysis

A bit of thought points up a fundamental weakness in the way that we have been thinking about separation. The permeate will generally have a very small concentration of the dissolved material. We are therefore handling a large volume of dilute solution by putting fresh permeate into each stage in the cross-flow configuration. A sensible approach would be to use the permeate leaving one stage as the feed to

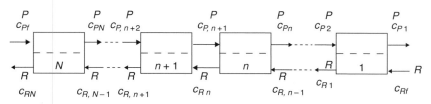

Figure 5.4. Schematic of countercurrent dialysis.

another. The driving force $c_R - c_P$ would not be as large in the second stage as in the first, but we would eliminate a considerable amount of fluid handling. The question then is how to connect the stages. The *countercurrent* configuration shown in Figure 5.4 is attractive, because it uses the most dilute permeate in the same stage as the most dilute raffinate (stage N), and the most concentrated permeate in the same stage as the most concentrated raffinate (stage 1), which should be the most efficient way to maintain a good driving force throughout the cascade.

We continue to assume that the amount of material transferred between the raffinate and permeate streams is sufficiently small that we can neglect any change in volumetric flow rates and treat R and P as constants from stage to stage, and we assume that the membrane area in each stage is the same. We then start by applying the steady-state balance equations to a typical stage n:

$$V_{R,n}\frac{dc_{R,n}}{dt} = 0 = R(c_{R,n-1} - c_{R,n}) - \Pi A(c_{R,n} - c_{P,n}), \qquad (5.21\text{a})$$

$$V_{P,n}\frac{dc_{P,n}}{dt} = 0 = P(c_{P,n+1} - c_{P,n}) + \Pi A(c_{R,n} - c_{P,n}). \qquad (5.21\text{b})$$

We also know that $c_{R0} = c_{Rf}$, and we will take c_{Pf}, which we rename $c_{P,N+1}$ for consistency, to be zero (pure permeate feed).

Equations 5.21a and 5.21b are a system of $2N$ linear algebraic equations in the variables $\{c_{R,n}\}$ and $\{c_{P,n}\}$, $n = 1, 2, \ldots, N$, with c_{R0} and $c_{P,N+1}$ as known constants. They can be solved in a variety of ways, including standard matrix methods. The traditional chemical engineering way to solve equations of this type, which arise frequently in separations applications, is to manipulate the equations to eliminate $c_{P,n}$, which, after some algebra, results in the following *second-order finite-difference equation*:

$$\left(\frac{\Pi A}{R} + 1\right)c_{R,n+1} - \left[2 + \left(\frac{R}{P} + 1\right)\frac{\Pi A}{R}\right]c_{R,n} + \left(\frac{\Pi A}{P} + 1\right)c_{R,n-1} = 0. \quad (5.22)$$

Linear finite-difference equations are solved by methods analogous to those for solving linear differential equations, as shown briefly in Appendix 5.A for those who have completed a course in differential equations. The solution, which can be verified by direct substitution into Equation 5.22, is

$$\frac{c_{R,n}}{c_{Rf}} = \frac{\frac{R}{P}\alpha^N - \alpha^n}{\frac{R}{P}\alpha^N - 1}, \qquad (5.23\text{a})$$

where

$$\alpha \equiv \frac{1 + \frac{\Pi A}{P}}{1 + \frac{\Pi A}{R}}.$$ (5.23b)

By setting $n = N$ we obtain the separation factor $s_R = c_{RN}/c_{Rf}$:

$$s_R = \frac{\frac{R}{P} - \left(\frac{1 + \frac{\Pi A}{P}}{1 + \frac{\Pi A}{R}}\right)^{-N}}{\frac{R}{P} - 1}.$$ (5.24)

We can analyze behavior using this result, but it is a bit easier to think again in terms of the continuous system by setting $A = A_T/N$ and letting N approach infinity. (In contrast to the cross-flow configuration, P remains unchanged.) Again, using the relation $(1 + \frac{x}{N})^N \to e^x$ as $N \to \infty$, we obtain

$$s_R = \frac{\frac{R}{P} - e^{-\left(\frac{\Pi}{P} - \frac{\Pi}{R}\right)A_T}}{\frac{R}{P} - 1}.$$ (5.25)

We then obtain the two useful limits that define countercurrent operation:

$$\left(\frac{P}{R}\right)_{\min} = \frac{s_R - 1}{s_R},$$ (5.26a)

$$A_{T,\min} = \frac{R}{\Pi} \ln(s_R + 1).$$ (5.26b)

Hence, we see that the minimum area requirement for the countercurrent configuration is not significantly different from that for the cross-flow. The smallest permeate rate possible with a large membrane area is very close to the raffinate rate. Actual operation will, of course, reflect some compromise between these limits. It is possible to repeat the optimization calculation done in Section 5.3 for the single-stage system, but in this case the analog of Equation 5.11 is a transcendental equation that does not have an analytical solution and would require numerical solution for specific values of the parameters.

5.7 Continuous Countercurrent Flow

The staged systems that we have considered up to this point are realistic for many separation processes, and the large N limit provides a convenient way to analyze continuous systems. Dialysis is generally carried out continuously, and it is instructive to address the continuous system directly. The process is shown schematically in Figure 5.5. The raffinate and permeate streams flow parallel to one another in two plane channels of length L that are separated by a membrane that has a width w. (The analysis is unchanged if the membrane is a tube, except that w becomes the perimeter of the tube.) The system is countercurrent if the flows are as shown, and that is the case that we consider here. Co-current operation is also possible.

Here we need to consider the choice of the control volume carefully, since, as noted in Section 4.2 and Figure 4.2, each characterizing variable must have a unique

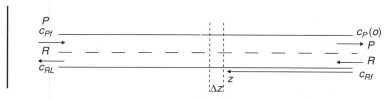

Figure 5.5. Schematic of continuous countercurrent dialysis.

value in the control volume. The concentration in each stream changes along the channel, so it is clear that the entire unit cannot be taken as the control volume. We choose instead a segment of each channel of length Δz, as shown in the figure; within this small region we can assume, as is typically done when using differential calculus, that the characterizing variables are constant, and we will subsequently let Δz go to zero. We assume, as before, that the amount of solute transferred between the streams is sufficiently small to allow us to take the flow rates R and P as constant in space. For this configuration we will start with the steady state in order to avoid the complication of having to deal with changes with respect to two independent variables, t and z. The steady-state mass balances in the two control volumes (raffinate side and permeate side) are then respectively

$$0 = R(c_R|_z - c_R|_{z+\Delta z}) - \Pi w \Delta z (c_R - c_P), \tag{5.27a}$$

$$0 = P(c_P|_{z+\Delta z} - c_R|_z) + \Pi w \Delta z (c_R - c_P). \tag{5.27b}$$

We now divide each equation by Δz and make use of the fact that the difference quotient $\frac{\varphi|_{z+\Delta z} - \varphi|_z}{\Delta z}$ goes to the derivative $d\varphi/dz$ as $\Delta z \to 0$. We therefore obtain the differential equations

$$R\frac{dc_R}{dz} + \Pi w (c_R - c_P) = 0, \tag{5.28a}$$

$$P\frac{dc_P}{dz} + \Pi w (c_R - c_P) = 0. \tag{5.28b}$$

Each equation requires one condition at a specific value of z in order to evaluate the constant of integration; these are $c_R(0) = c_{Rf}$ and $c_P(L) = c_{Pf} = 0$.

By subtracting Equation 5.28b from Equation 5.28a we obtain $R\frac{dc_R}{dz} - P\frac{dc_P}{dz} = 0$, which integrates to $Rc_R(z) - Pc_p(z) = \text{constant} = Rc_{RL}$. (Note that c_{RL} is still unknown.) We can then substitute into Equation 5.28a to obtain a single equation for the raffinate concentration, as follows:

$$\frac{dc_R}{dz} = c_R \left(\frac{\Pi w}{P} - \frac{\Pi w}{R} \right) - \frac{\Pi w}{P} c_{RL}. \tag{5.29}$$

We have seen this type of equation before; it is of the form of Equation 4.11, for example. Hence, we can write the solution with $c_R(0) = c_{Rf}$ as

$$c_R(z) = c_{Rf} e^{\left(\frac{\Pi w}{P} - \frac{\Pi w}{R} \right)z} + \frac{c_{RL}}{1 - \frac{P}{R}} \left[1 - e^{\left(\frac{\Pi w}{P} - \frac{\Pi w}{R} \right)L} \right]. \tag{5.30}$$

We now evaluate the unknown c_{RL} by setting $z = L$ in Equation 5.30. It is readily established that the resulting equation for $s_R = c_{Rf}/c_{RL}$ is identical to Equation 5.25, with $A_T = wL$. This result is, of course, expected.

5.8 Hemodialysis

In hemodialysis, the circulatory system of the patient is connected directly to the dialyzer through shunts in the arm. The shunts are typically installed permanently, since the patient must be able to connect easily to the system three or more times a week. Truskey and coworkers have compiled data on hemodialysis from a variety of sources, and they have computed the various contributions to Π for urea passing through a 180-μm-diameter cellulosic hollow fiber membrane with a wall thickness of 20 μm from membrane properties, the physical properties of urea, and typical flow conditions. The values that they compute are consistent with those reported for commercial dialyzers. We obtain $\Pi = 0.33$ cm/min from their values for the various contributions to the resistance, where 40 percent of the resistance comes from the membrane permeability. The typical operating parameters that they report are $R = 250$ cm^3/min, $P = 500$ cm^3/min, and $A_T = 1,200$ cm^2. (Truskey and colleagues report typical values of ΠA_T for urea to be in the range 300–500 cm^3/min, from which we obtain the 1,200 cm^2 value for the typical area.) Substituting these values into Equation 5.25 gives a separation factor $s_R = c_{RL}/c_{Rf}$ of about 3.5 for urea. The minimum surface area computed from Equation 5.26b for this separation factor is 1,130 cm^2, so the actual area in this example is only about 6 percent greater than the minimum.

5.9 Concluding Remarks

Membrane separation is an illustrative and very important example of the use of the conservation equations for nonreacting systems and the connection between analysis and design. In a broader context, however, the various dialysis configurations illustrate a very important point regarding the marginal cost of improved separation that is common to all separation processes. Each additional factor of ten in separation requires a membrane area equal to that required for the previous factor of ten. Hence, there is an exponential increase in size, with a corresponding exponential increase in capital cost, for increased purity.

The large marginal capital cost of improved separations, which is especially important in large industrial processes, brings up the obvious issue of the trade-off between cost and purity. Simply stated, when is the extra decimal place needed? Is there a clear economic or societal gain in going from 99 percent purity to 99.9 percent that justifies doubling the size and capital cost? Would the money be better invested in purifying another stream to 90 percent (or even to 50%)? Indeed, in the context of environmental control, this simple calculation (in reality, the equivalent calculations for the specific separations processes of interest) provides some of the intellectual underpinning for the controversial concept of "trading" of pollution rights, wherein organizations, or different effluent sources within a single organization, can effectively pool the contents of their effluents to achieve an overall goal, rather than requiring that each individual source meet the goal. Emissions trading for SO_2 control to reduce "acid rain" is incorporated, for example, in the 1990 U.S. Clean Air Act.

Bibliographical Notes

A comprehensive treatment of membrane science and technology can be found in

> Baker, R. W., *Membrane Technology and Applications*, Wiley, New York, 2004.

The hemodialysis data are from

> Truskey, G. A., F. Yuan, and D. F. Katz, *Transport Phenomena in Biological Systems*, Pearson Prentice Hall, Upper Saddle River, NJ, 2004, pp. 375–384.

The calculation of maxima and minima for optimization problems without constraints is a topic that is usually covered in introductory calculus courses, and constrained optimization is usually covered in advanced calculus. Most books on optimization methods for engineers are designed for advanced study. The first two chapters of

> Denn, M. M., *Optimization by Variational Methods*, McGraw-Hill, New York, 1969; corrected reprint edition Robert Krieger, Huntington, NY, 1978

address both analytical and computational methods for problems of the type considered in this chapter and elsewhere in this text at a level that requires only a first course in differential calculus. The book is out of print but widely available.

There is a very good discussion of trading pollution rights in

> Beder, S., "Trading the earth: The politics behind tradeable pollution rights," *Environmental Liability*, **9**, 152–160 (2001).

Beder is an Australian engineer who has become a social scientist. The paper was available electronically at the time of writing at http://works.bepress.com/sbeder/11/. This citation provides an important cautionary note about using Web sites as references, since this is not the site at which the paper was posted when this chapter was first drafted; that site disappeared in the interim.

PROBLEMS

5.1. A dialysis system is set up to reduce the concentration of a solute from 100 g/L to 1 g/L. The membrane permeability is 0.001 m/s, and the raffinate flow rate is 0.25 L/min. The permeate stream enters with none of the solute present.

    a.   What is the minimum permeate flow rate?

    b.   What is the minimum membrane area required?

5.2. Derive Equation 5.3 directly by doing a steady-state mass balance with a control volume that includes *both* V_R and V_P.

5.3. Consider a dialysis system in which the membrane has a permeability of $1/3$ cm/min and the total available membrane area is 2,700 cm². The raffinate feed rate is 300 mL/min and the total permeate flow available is 900 mL/min. Compare the separation factors for the following cases: (a) a single stage; (b) a nine-stage cross-flow configuration; (c) a nine-stage countercurrent configuration.

5.4. Consider a continuous countercurrent dialysis system that uses a flat membrane with a width of 25 cm and a permeability of $1/3$ cm/min. The raffinate and permeate feeds are 300 mL/min and 900 mL/min, respectively. The required separation factor is 7.24. What is the required length? Discuss the result in the context of your solution to part c of problem 5.3.

5.5. Starting with Equation 5.30, carry out the steps that lead to the separation factor given by Equation 5.25.

5.6. Repeat the optimization analysis in Section 5.3 for a continuous countercurrent dialysis process, but do not attempt to solve the equation that you derive for the optimal permeate flow rate.

Appendix 5A: Linear Finite-Difference Equations

A homogeneous linear finite-difference equation has the form

$$a_n x_n + a_{n-1} x_{n-1} + \cdots + a_1 x_1 + a_0 = 0, \tag{5A.1}$$

where the coefficients $\{a_i\}$ are constants. It can be shown by direct substitution that the solution is of the form

$$x_k = C_1 m_1^k + C_2 m_2^k + \cdots C_n m_n^k, \tag{5A.2}$$

where the $\{m_i\}$ are the n roots of the nth order polynomial equation

$$a_n m^n + a_{n-1} m^{n-1} + \cdots + a_1 m + a_0 = 0. \tag{5A.3}$$

This result is analogous to the solution of linear homogeneous ordinary differential equations, where the solution is a sum of exponentials. The n coefficients $\{C_i\}$ must be evaluated from n independent conditions. Nonhomogeneous linear difference equations also have particular solutions, which are found in a manner analogous to the particular solutions to nonhomogeneous linear ordinary differential equations.

Appendix 5B: An Optimization Calculation

Consider a function $\Im(x_1, x_2, \ldots, x_N) = \prod_{n=1}^{N} f(x_n)$. The cross-flow staged membrane separation example in Equation 5.20 is a special case of this general form. We wish to choose the N variables $\{x_n\}$ that maximize $\Im$ subject to the constraint $\sum_{n=1}^{N} x_n = X$. The problem is made a bit simpler by noting that the variables that maximize $\Im$ will also maximize $\ln \Im$; hence, we may reformulate the problem as follows:

$$\text{Maximize } \ln \Im(x_1, x_2, \ldots, x_N) = \sum_{n=1}^{N} \ln f(x_n) \text{ subject to } \sum_{n=1}^{N} x_n = X. \quad (5B.1)$$

There are now two approaches that we can take. For those familiar with the use of Lagrange multipliers for constrained maximization or minimization, the process is straightforward. We differentiate the function $\sum_{n=1}^{N} \ln f(x_n) + \lambda(\sum_{n=1}^{N} x_n - X)$ with respect to each element of the set $\{x_n\}$ and set the result to zero. This gives N equations that, together with the constraint $\sum_{n=1}^{N} x_n = X$, provide $N + 1$ equations to find the $N + 1$ variables $x_1, x_2, \ldots, x_N, \lambda$. Thus, we have

$$\frac{\partial}{\partial x_k}\left[\sum_{n=1}^{N} \ln f(x_n) + \lambda\left(\sum_{n=1}^{N} x_n - X\right)\right] = \frac{f'(x_k)}{f(x_k)} + \lambda = 0, \quad k = 1, 2 \ldots, N. \quad (5B.2)$$

Here, $f'(x_k) = df(x_k)/dx_k$. The ratio $f'(x_k)/f(x_k)$ is a known function of the variable x_k. It therefore follows immediately that all of the N variables $\{x_n\}$ independently satisfy exactly the same algebraic equation; hence, it must be true that $x_1 = x_2 = \cdots = x_N = \frac{X}{N}$.

The process is a bit more involved for those who are not familiar with the use of Lagrange multipliers, which are often introduced in the mathematics curriculum only in courses in advanced calculus. In that case, we need to convert the constrained problem to an unconstrained one, which is straightforward for this example but not so in general. We are free to choose only $N - 1$ of the N variables $\{x_n\}$ independently; the N-th is determined by the constraint. Hence, we write $x_N = X - \sum_{n=1}^{N-1} x_n$. The problem to be solved is now

$$\text{Maximize } \ln \Im(x_1, x_2, \ldots, x_{N-1}) = \sum_{n=1}^{N-1} \ln f(x_n) + \ln f\left(X - \sum_{n=1}^{N-1} x_n\right). \quad (5B.3)$$

This is an *unconstrained* maximization problem, so we find the solution by differentiating the function $\ln \Im$ with respect to each of its $N - 1$ arguments in turn and setting the result to zero. We thus obtain

$$\frac{\partial}{\partial x_k}\left[\sum_{n=1}^{N-1} \ln f(x_n) + \ln f\left(X - \sum_{n=1}^{N-1} x_n\right)\right] = \frac{f'(x_k)}{f(x_k)} + (-1)\frac{f'\left(X - \sum_{n=1}^{N-1} x_n\right)}{f\left(X - \sum_{n=1}^{N-1} x_n\right)} = 0,$$

$$k = 1, 2, \ldots, N - 1. \quad (5B.4)$$

We can rewrite this equation as

$$\frac{f'(x_k)}{f(x_k)} = \frac{f'\left(X - \sum_{n=1}^{N-1} x_n\right)}{f\left(X - \sum_{n=1}^{N-1} x_n\right)} = \frac{f'(x_N)}{f(x_N)}, \quad k = 1, 2, \ldots N - 1. \quad (5B.5)$$

Hence, it follows that each $x_k = x_N$ for all $k = 1, 2, \ldots, N - 1$; that is, $x_1 = x_2 = \cdots = x_N = \frac{X}{N}$.

Chemically Reacting Systems

6.1 Introduction

Most chemical engineering applications involve chemical reactions; this is true whether we are dealing with the manufacture of computer chips, the creation of scaffolding for cell growth in artificial organs, the design of a novel battery, or the conversion of biomass to synthetic fuel. Chemical reactions can take place in gas, liquid, or solid phases; in the bulk or at interfaces; and with or without catalysts. Many reactions of social, physiological, or industrial significance take place in multiphase environments, where reactive species and reaction products must cross phase boundaries. Some "cracking" reactions for the production of intermediate molecular weight hydrocarbons from heavy components of crude oil, for example, are carried out in "trickle-bed" reactors; here, liquid and gas phases flow together over a bed of catalyst particles. Design and operating considerations for trickle-bed reactors include ensuring the necessary contact between the phases so that chemical species can get to where they need to be for specific reactions to occur. Similarly, many biochemical reactions, including wastewater treatment, require that suspended microorganisms be able to access organic nutrients that are dissolved in the liquid phase, as well as oxygen that is supplied as a gas (perhaps in air) and must dissolve into the liquid.

The design and operating issues for complex reactors of the types mentioned here are addressed in advanced courses, but we can focus on some basic principles essential to our overall understanding that can be elucidated by considering single-phase reactors and simple geometries; many important reactions are, in fact, carried out in a single phase in relatively simple geometries. There are two basic designs, stirred tank and tubular. The terms are descriptive: The stirred-tank reactor is typically cylindrical, with a height-to-diameter ratio of no more than two or three, so that effective mixing can take place. There are often baffles to enhance mixing. A schematic of a typical continuous-flow stirred-tank reactor (CFSTR, or sometimes CSTR or C*) is shown in Figure 6.1; the reactants enter through one or more streams at the top and the products leave through a stream at the bottom in this design. There is a jacket around the outside of the reactor through which a liquid or steam flows for temperature control, a topic that we discuss in Chapter 13.

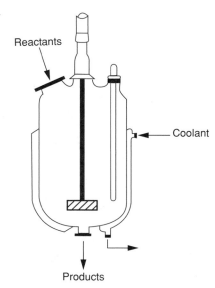

Figure 6.1. Schematic of a typical continuous-flow stirred-tank reactor.

A tubular reactor, as the name suggests, is constructed from a long tube, so longitudinal mixing is negligible, although the system is often assumed to be well mixed in the radial direction. It is quite common for the reactor tube to be coiled, or to be folded back on itself many times, in order to fit within a confined space, and the tube is typically surrounded by a jacket for heating or cooling. We will not consider tubular reactors in this introductory treatment, but the analysis is straightforward and follows along the lines of the treatment of the continuous dialyzer in Section 5.7.

The mass balance equations for a chemically reacting system are the same as for the other systems that we have analyzed thus far, with one important distinction: The masses of individual species in the control volume can now change because of creation or destruction by chemical reaction. In general, chemical reaction is a volumetric process (except for surface reactions, of course); hence, all other things being equal, the total rate of creation or destruction of a species by chemical reaction in units of mass/time will be proportional to the available volume. We therefore express rates on a volumetric basis, in units of mass/(time × volume); in fact, since chemical reactions take place between *moles* of the reactant species, the best way to keep track of the mass of any species is to employ molar units. We will use the generic symbol r to denote a rate at which a reaction takes place, with units of moles/(time × volume); the rates of formation or disappearance of individual species by chemical reaction will be indicated by appropriate subscripts.

6.2 Continuous-Flow Stirred-Tank Reactor

Consider the reactor shown schematically in Figure 6.2. We have a well-stirred tank in which a chemical reaction occurs; the reaction is as follows:

$$\alpha A + \beta B + \cdots \rightarrow \mu M + \nu N + \cdots. \tag{6.1}$$

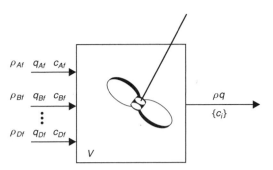

Figure 6.2. Schematic of a continuous-flow stirred-tank reactor.

Here, A, B,...represent the reactants, and M, N,...represent the reaction products. α, β, ... are the stoichiometric coefficients of the reactants, and μ, ν, ... are the stoichiometric coefficients of the products; that is, α moles of A combine with β moles of B, and so on, to form μ moles of M, ν moles of N, and so on. There is never a loss of generality in taking $\alpha = 1$, since we can always divide Equation 6.1 by α; in that case the stoichiometric coefficients may not be integers, but they will be ratios of integers except in unusual cases. The reactants might enter in separate streams, as shown, or two or more components of the feed might enter together, in which case $\rho_{Af}q_{Af} + \rho_{Bf}q_{Bf} + \cdots = \rho_f q_f$.

For definiteness, we will consider the irreversible reaction $A + \beta B \to \mu M$. The generalization to a single reaction with more reactants or products and to a reversible reaction will be obvious. The reactor is the control volume and the overall mass balance is unchanged:

$$\frac{d}{dt}\rho V = \rho_{Af}q_{Af} + \rho_{Bf}q_{Bf} - \rho q. \tag{6.2}$$

We need to include the reaction rate in the mass balance equation for each species. We will denote the rate at which species i is created by chemical reaction, in units of moles/(time $\times$ volume), by r_{i+}, and the rate at which species i is destroyed by chemical reaction as r_{i-}. The reaction rates r_{i+} and r_{i-} are always positive numbers. (Note that in our example here, A and B are only destroyed and M is only created.) The species balance equations are then

$$\frac{d}{dt}c_A V = c_{Af}q_{Af} - c_A q - r_{A-}V, \tag{6.3a}$$

$$\frac{d}{dt}c_B V = c_{Bf}q_{Bf} - c_B q - r_{B-}V, \tag{6.3b}$$

$$\frac{d}{dt}c_M V = -c_M q + r_{M+}V. \tag{6.3c}$$

The last term on the right side of each equation is the rate of disappearance or formation of the relevant species by reaction. Concentrations are in units of moles/volume, and we have assumed for simplicity that the product M is not contained in any feedstream.

Now, every time that one mole of A is reacted, β moles of B are also reacted and μ moles of M are formed. Hence, the rates are not independent. It follows immediately that $r_{A-} = r_{B-}/\beta = r_{M+}/\mu$, and only a single rate is required; we will choose this to be r_{A-}, which we will simply denote r. r is frequently called the *intrinsic rate of reaction*. Equations 6.3a,b, and c then become

$$\frac{d}{dt}c_A V = c_{Af}q_{Af} - c_A q - rV,$$ (6.4a)

$$\frac{d}{dt}c_B V = c_{Bf}q_{Af} - c_B q - \beta rV,$$ (6.4b)

$$\frac{d}{dt}c_M V = -c_M q + \mu rV.$$ (6.4c)

We saw in Chapter 4 that the overall mass balance, Equation 6.2, was independent of the densities of the individual streams provided that the mixture was linear in concentration, which is equivalent to there being no volume change on mixing. We might expect this condition to become more restrictive when there is chemical reaction, and in fact there is now a second condition: The coefficients in the linear expression for the density must be proportional to the molecular weights of the corresponding species. This restriction, which is derived in Appendix 6A, will often be satisfied, and we can then write the overall mass balance as

$$\frac{dV}{dt} = q_{Af} + q_{Bf} - q.$$ (6.5)

The reaction rate, r, clearly depends on the composition. (r must equal zero if there is no A or no B in the reactor.) Furthermore, the composition dependence will certainly be different for different reacting systems. Hence, we require a constitutive equation for the intrinsic rate $r(c_A, c_B, \dots)$ in order to complete the description.

6.3 Reaction Kinetics

The traditional picture of a chemical reaction is that reactant molecules collide and form the product. A more sophisticated version is that only a certain fraction of collisions actually result in reaction. In either case, the rate of a reaction will be proportional to the probability that a collision of reactants occurs in a unit time. The probability that two molecules collide will depend on the concentration of each, and will be equal to zero if the concentration of either species is zero; the simplest case assumes that the probability is proportional to the concentration of each reactant. For a case in which one molecule of A must react with β molecules of B, the rate will then be of the form

$$r = kc_A c_B^{\beta}.$$ (6.6)

This form of a rate expression is commonly known as *mass action kinetics*. The *rate constant* k will usually be a function of the temperature, and perhaps also of the pressure in a gas-phase system. It is often the case that the functional form is more complex, perhaps because there are intermediate reactions occurring that do not show up in the overall kinetic scheme. *Product inhibition* occurs in some catalytic and biological reactions, and the concentration of product may appear in the rate expression. Power functions are in common use, in large measure because they are adequate to describe complex functions over limited ranges of the variables, and the exponents may in some cases be different from the stoichiometric coefficients. The study of reaction mechanisms and the corresponding rate expressions is an important part of the subject of *chemical kinetics*, which is usually incorporated in the chemical engineering curriculum in a core course with a name such as "Reaction Kinetics and Reactor Design." For our purposes in this introductory text, we will consider the reaction rates to be empirical functionalities that we can obtain by experiment, and we will not delve into the underlying molecular processes.

6.4 Batch Reactor

A batch reactor is a well-stirred tank that has no inflow or outflow. Batch reactors are used in many manufacturing processes, especially in the pharmaceutical and fine chemicals industries, but our interest in them here is that they are ideal configurations for the determination of reaction rate constitutive relations. For the latter application the reactors are usually laboratory-scale setups, often nothing more than a beaker with a stirrer. As we shall see, the challenge in using a batch reactor for rate determination is the ability to obtain good conversion data as a function of time.

We will continue to consider the irreversible reaction $A + \beta B \to \mu M$ for definiteness, although the generality of the approach will be obvious. Without flow, the volume V is a constant ($dV/dt = 0$), and Equations 6.4a,b, and c become

$$\frac{dc_A}{dt} = -r, \quad \frac{dc_B}{dt} = -\beta r, \quad \frac{dc_M}{dt} = +\mu r. \qquad (6.7a,b,c)$$

It follows immediately that $dc_B/dt = \beta dc_A/dt$ and $dc_M/dt = -\mu dc_A/dt$. Two relations then follow by integration:

$$c_B(t) - c_{B0} = \beta[c_A(t) - c_{A0}], \qquad (6.8a)$$

$$c_M(t) - c_{M0} = -\mu[c_A(t) - c_{A0}]. \qquad (6.8b)$$

c_{A0}, c_{B0}, and c_{M0} are the molar concentrations in the reactor at time $t = 0$, with c_{M0} typically equal to zero. Thus, we need solve for only one concentration. We will assume for illustration that that concentration is A, and the equation that we need to solve is Equation 6.7a.

6.5 A + B → μM

We now consider the irreversible chemical reaction in which one mole of A reacts with one mole of B to form μ moles of M (i.e., the case considered above, but with

Table 6.1. *Concentration of H_2SO_4 versus time for the reaction of sulfuric acid with diethyl sulfate in aqueous solution at 22.9°C. Data of Hellin and Jungers,* Bull. Soc. Chim. France, *No. 2, pp. 386–400 (1957).*

Time, t (min)	Concentration of H_2SO_4, $c_A(t)$ (g-mol/L)
0	5.50
41	4.91
48	4.81
55	4.69
75	4.38
96	4.12
127	3.84
162	3.59
180	3.44
194	3.34

$\beta = 1$). We will assume mass action kinetics, so $r = kc_Ac_B$. With Equation 6.8a we can then write

$$\frac{dc_A}{dt} = -kc_Ac_B = -kc_A(c_A + c_{B0} - c_{A0}) = -kc_A(c_A + \psi), \qquad (6.9)$$

where $\psi \equiv c_{B0} - c_{A0}$. We assume that the temperature in the batch reactor does not change with time, so k is a constant. The integration of Equation 6.9 is straightforward. We rewrite the equation in *separated* form as

$$\int\limits_{c=c_{A0}}^{c_A(t)} \frac{dc}{c(c + \psi)} = -kt, \qquad (6.10)$$

where c is the dummy variable of integration. There are two cases that need to be considered separately, $\psi = 0$ ($c_{A0} = c_{B0}$) and $\psi \neq 0$ ($c_{A0} \neq c_{B0}$). The integration of Equation 6.10 is straightforward in either case, with the following results:

$$\psi = 0 : \frac{1}{c_A(t)} = \frac{1}{c_{A0}} + kt, \qquad (6.11a)$$

$$\psi \neq 0 : \ln\left[\frac{c_A(t) + \psi}{c_A(t)}\right] - \ln\left[\frac{c_{A0} + \psi}{c_{A0}}\right] = \ln\left[\frac{c_B(t)}{c_A(t)}\right] - \ln\left[\frac{c_{B0}}{c_{A0}}\right] = \psi kt. \qquad (6.11b)$$

Thus, if the initial concentrations of the two reactants are equal, a plot of $1/c_A$ versus t will yield a straight line with slope k. Similarly, for unequal initial concentrations, Equation 6.11b indicates that a plot of a slightly more complicated function will give a straight line with slope ψk. (It is easily shown that Equation 6.11b reduces to Equation 6.11a in the limit $\psi \to 0$ by making use of the relation $\ln(1 + x) \approx x$ for $x \ll 1$. See Problem 6.6.)

EXAMPLE 6.1 The data in Table 6.1 were obtained for the reaction of sulfuric acid with diethyl sulfate, $H_2SO_4 + (C_2H_5)_2SO_4 \to 2C_2H_5SO_4H$, in aqueous solution at 22.9°C. The initial concentrations of both reactants were the same, 5.5 g-mol/L. What is the rate expression?

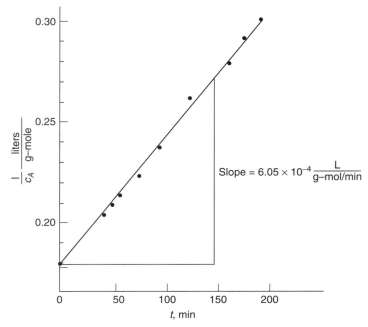

Figure 6.3. Computation of the second-order rate constant for the reaction between sulfuric acid and diethyl sulfate in aqueous solution.

In our symbolic nomenclature, H_2SO_4 is denoted by A, $(C_2H_5)_2SO_4$ is denoted by B, and $2C_2H_5SO_4H$ is denoted by M. $c_{A0} = c_{B0} = 5.5$ g-mol/L. $\mu = 2$ and $\psi = 0$. We assume mass action kinetics, so the rate with equal initial concentrations will be of the form $r = kc_A^2$.[*] As required by Eq. 6.11a, the data are plotted in Figure 6.3 as $1/c_A$ versus t. There is some experimental scatter, but the data are clearly represented by the straight line shown in the figure. The rate constant is then determined from the slope to be $k = 6.05 \times 10^{-4}$ L/(g-mol min).

EXAMPLE 6.2 The data in Table 6.2 were taken in a batch reactor for the reaction between sodium ethoxide and ethyl dimethyl sulfonium iodide in solution in ethanol:

$$
\text{NaOC}_2\text{H}_5 + \text{C}_2\text{H}_5-\!\!\overset{\displaystyle \text{CH}_3}{\underset{\displaystyle \text{CH}_3}{\text{S}}}\!\!-\!\text{I} \longrightarrow \text{products}
\quad
\begin{bmatrix}
\text{NaI} + (\text{C}_2\text{H}_5)_2\text{O} + \text{S}(\text{CH}_3)_2 \\
\text{or} \\
\text{NaI} + \text{C}_2\text{H}_5\text{OH} + \text{C}_2\text{H}_4 + \text{S}(\text{CH}_3)_2
\end{bmatrix}.
$$

Verify that the data are consistent with mass action kinetics and find the rate constant.

We denote sodium ethoxide by A and ethyl dimethyl sulfonium iodide by B. According to Equation 6.11b, a plot of $\ln[c_B(t)/c_A(t)]$ versus t should give

[*] Problem 6.1 asks you to analyze this data set using the methodology of Appendix 2A to show that $n = 2$ is a reasonable exponent for these data if the rate expression is assumed to be of the form $r = kc_A^n$.

Table 6.2. *Concentrations of sodium ethoxide and ethyl dimethyl sulfonium iodide in ethanol solution as functions of time. Data of E. D. Hughes, C. K. Ingold, and G. A. Maw, J. Chem. Soc., 2072–2077 (Dec., 1948).*

Time (min)	Concentration of $NaOC_2H_5$, c_A (g-mol/L)	Concentration of $C_2H_5S(CH_3)_2I$, c_B (g-mol/L)	$c_B - c_A$ (g-mol/L)
0	0.0961	0.0472	− 0.0489
12	0.0857	0.0387	− 0.0470
20	0.0805	0.033	− 0.0471
30	0.0749	0.0278	− 0.0471
42	0.0698	0.0228	− 0.0470
51	0.0671	0.0200	− 0.0471
63	0.0638	0.0168	− 0.0470
∞	0.0470	0	− 0.0470

a straight line for mass action kinetics. Note from Table 6.2 that $\psi = -0.0470$ g-mol/L, and that there seems to be some discrepancy in the measurement of the initial concentration of c_A. The data are plotted in Figure 6.4, and it is evident that they do follow a straight line, with the exception of the intercept point. The mass action rate expression is thus verified, and we obtain the rate constant k from the relation $k = \text{slope}/\psi = (-0.0103)/(-0.0470) = 0.21$ L/(g-mol min). It is interesting to note how different this rate constant is from the one in the previous example. Rate constants for different reactions can vary by many orders of magnitude.

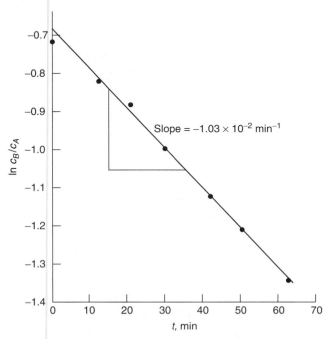

Figure 6.4. Computation of the second-order rate constant for the reaction between sodium ethoxide and ethyl dimthyl sulfonium iodide in ethanol.

Table 6.3. *Data for the decomposition of NDEA with ultraviolet radiation, read from a graph in B. Xu, Z. Chen, F. Qi, J. Ma, and F. Wu, J. Hazardous Materials, **179**, 976–982 (2010).*

Time (min)	c_{NDEA}/c_{NDEA0}
0	1.0
2	0.31
5	0.053
10	too small to read from graph

6.6 First-Order and Pseudo-First-Order Reactions

Reaction rates sometimes appear to be proportional to the concentration of a single species, and these are called *first-order* reactions. Spontaneous decomposition, as occurs in nuclear fission, would be expected to be truly first order, but other reactions may exhibit first-order behavior for a variety of reasons. Chemical reactions of the form A + B $\rightarrow$ μM in solution may sometimes be carried out with c_B much larger than c_A, for example, perhaps because a large excess of B is required to suppress an unwanted side reaction. The role of B is effectively masked in the mathematical description of the reaction in such a case, and the resulting equations are somewhat simpler. Consider Equation 6.9: If $c_{B0} \gg c_{A0}$, then $c_{B0} \gg c_A(t) - c_{A0}$ for all t. Hence, to a good approximation, we may write Equation 6.9 as

$$\frac{dc_A}{dt} \cong -(kc_{B0})c_A = -k'c_A, \tag{6.12}$$

where $k' = kc_{B0}$ has units of inverse time, say min^{-1}, and the kinetics appear to be first order. k' is often called a *pseudo-first-order* rate constant, and the reaction a pseudo-first-order reaction. Solution of Equation 6.12 leads to

$$\ln \frac{c_A(t)}{c_{A0}} = -k't, \tag{6.13}$$

so a plot of the logarithm of c_A versus t will be linear, with a slope equal to $-k'$. (This result can also be obtained directly from Equation 6.11b by setting $c_B(t) = c_{B0}$.)

Another way in which a reaction in solution might appear to be first order is if a second species is required for a reaction to occur, but the second species, or *homogeneous catalyst*, is not used up in the reaction. That is, the chemical equation will have the form A + B $\rightarrow$ μM + B, and hence will appear to be of the form A $\rightarrow$ μM. The concentration of B is then unchanged, and Equation 6.12 is an exact representation.

EXAMPLE 6.3 N-nitrosodiethylamine (NDEA) is a carcinogenic compound that is found in drinking water, together with other nitrosamines. One way in which NDEA can be broken down is by exposure to ultraviolet radiation. The data in Table 6.3 were read from a published graph reporting on an experiment in

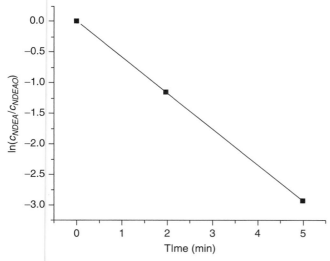

Figure 6.5. Logarithm of relative NDEA concentration as a function of time.

which 0.10 mmol/L of NDEA at pH 6.0 was exposed to 1 mW/cm² of ultraviolet radiation. Are the data consistent with first-order kinetics?

The data are plotted as $\ln(c_{NDEA}/c_{NDEA0})$ versus time in Figure 6.5. The available data points lie on a straight line passing through the origin with a slope of -0.59 min^{-1}. Hence, according to Equation 6.13, the limited available data are consistent with first-order or pseudo-first-order kinetics.

EXAMPLE 6.4 The cleavage of diacetone alcohol in the presence of hydroxide ion (e.g., aqueous NaOH) at 25°C appears to be a simple first-order decomposition:

$$CH_3-\underset{\underset{OH}{|}}{\overset{\overset{CH_3}{|}}{C}}-CH_2-\overset{\overset{O}{\|}}{C}-CH_3 \rightarrow 2CH_3-\overset{\overset{O}{\|}}{C}-CH_3.$$

The reaction is actually catalyzed by the hydroxide, and the true reaction is

$$CH_3-\underset{\underset{OH}{|}}{\overset{\overset{CH_3}{|}}{C}}-CH_2-\overset{\overset{O}{\|}}{C}-CH_3 + OH^- \rightarrow 2CH_3-\overset{\overset{O}{\|}}{C}-CH_3 + OH^-.$$

Data for the apparent first-order reaction-rate constant are shown in the first two columns of Table 6.4 for various concentrations of NaOH in the reaction mixture. Show that the reaction exhibits mass action kinetics with a true rate $r = kc_A c_B$.

The third column in the table shows the pseudo-first-order rate constant k' divided by the concentration of NaOH. The result is a constant value to within

Table 6.4. *Pseudo-first-order and actual second-order rate constants as functions of NaOH concentration for the homogeneously catalyzed cleavage of diacetone alcohol at 25°C. Data of C. C. French, J. Am. Chem. Soc., 51, 3215–3225 (1929).*

NaOH (c_B) (g-mol/L)	k' (min^{-1})	$k = k'/c_B$ (L/g-mol min)
5×10^{-3}	2.32×10^{-3}	0.465
10×10^{-3}	4.67×10^{-3}	0.467
20×10^{-3}	9.40×10^{-3}	0.470
40×10^{-3}	$19.2 \ \times 10^{-3}$	0.479
100×10^{-3}	$47.9 \ \times 10^{-3}$	0.479

3 percent, indicating that the true reaction rate is $r = kc_A c_B$, where $k = 0.47$ L/(g-mol min).

6.7 Reversible Reactions

All chemical reactions are reversible in principle, and most are reversible in practice as well. Hence, we must now return to the reaction in Equation 6.1 and consider the more general case of reversibility. As before, we take $\alpha = 1$ without loss of generality, and all other stoichiometric coefficients other than β and μ to be zero. We thus have two reactions to consider:

$$A + \beta B \rightarrow \mu D$$

$$\mu D \rightarrow A + \beta B$$

We will therefore have to account for both the forward and reverse reactions in the species conservation equations. In the batch reactor, with the usual assumptions that lead to constant volume, the species equations then become

$$\frac{dc_A}{dt} = r_{A+} - r_{A-}, \tag{6.14a}$$

$$\frac{dc_B}{dt} = r_{B+} - r_{B-}, \tag{6.14b}$$

$$\frac{dc_M}{dt} = r_{M+} - r_{M-}. \tag{6.14c}$$

All of the stoichiometric arguments used previously still apply. Clearly, β moles of B still vanish for every mole of A that is reacted, and, similarly, β moles of B are formed for every mole of A that is former by the reverse reaction. μ moles of M are formed for every mole of A that is reacted, and one mole of A and β moles of B are formed for every μ moles of M that are lost in the reverse reaction. Hence,

$r_{B-} = \beta r_{A-}, r_{M+} = \mu r_{A-}, r_{B+} = \beta r_{A+}$, and $r_{M-} = \mu r_{A+}$. We can then again define a single reaction rate, denoted r, as follows:

$$r \equiv r_{A-} - r_{A+} = \frac{r_{B-} - r_{B+}}{\beta} = \frac{r_{M+} - r_{M-}}{\mu}. \tag{6.15}$$

We thus recover Equations 6.6a, b, and c, which clearly depend only on the fact that there is a single reaction, and not on the assumption if irreversibility. This is also true of Equations 6.7a and b, which relate the concentrations of the various reactants and products in the batch reactor at any time. Note that r is the difference between two positive quantities and may be positive or negative.

We can gain considerable insight by considering the specific example of $\beta = 1$ and $\mu = 2$, together with mass action kinetics. We thus write r_{A-} (the forward rate) as $k_1 c_A c_B$ and r_{A+} (the reverse rate) as $k_2 c_M^2$, with $r = k_1 c_A c_B - k_2 c_M^2$. The batch reactor equations are thus

$$\frac{dc_A}{dt} = \frac{1}{\beta}\frac{dc_B}{dt} = -\frac{1}{\mu}\frac{dc_M}{dt} = -k_1 c_A c_B + k_2 c_M^2. \tag{6.16}$$

Note that all rates of change of concentrations with time are zero if $k_1 c_A c_B = k_2 c_M^2$; that is, *the system is at equilibrium*, which simply means that forward rates exactly equal reverse rates. We denote the equilibrium concentrations of the species with a subscript e; we therefore obtain

$$\frac{c_{Me}^2}{c_{Ae} c_{Be}} = \frac{k_1}{k_2} = K_{eq}. \tag{6.17}$$

K_{eq} is, of course, the *equilibrium constant*, which is familiar from the general chemistry course.

We can make one observation that is of experimental significance even before we complete the mathematical steps to determine the time evolution of the concentrations in a batch reactor. Suppose that we begin the experiment with no M present. In that case, c_M will be very small, and the second term on the right of Equation 6.16, which is quadratic in c_M, will be negligible for a finite period of time. The equation will then be approximately the same as Equation 6.9, and the system will appear to be irreversible. This is another example of the significance of the time scale when evaluating system response.

For specificity we consider the case in which $\alpha = \beta = 1$ and $\mu = 2$, with $c_{A0} = c_{B0}$ and $c_{M0} = 0$. It then follows from Equation 6.8 that $c_A(t) = c_B(t)$ and $c_M(t) = 2[c_{A0} - c_A(t)]$. Equation 6.16 can then be written as

$$\frac{dc_A}{dt} = -k_1 c_A^2 + k_2 [2(c_{A0} - c_A)]^2, \tag{6.18}$$

and the equilibrium relation is

$$K_{eq} = \frac{k_1}{k_2} = 4\left(\frac{c_{A0}}{c_{Ae}} - 1\right)^2. \tag{6.19}$$

Table 6.5. *Concentration of H_2SO_4 versus time for the reaction of sulfuric acid with diethyl sulfate in aqueous solution at 22.9°C, full data set. Data of Hellin and Jungers, Bull. Soc. Chim. France, No. 2, pp. 386–400 (1957).*

Time, t (min)	concentration of H_2SO_4, $c_A(t)$ (g-mol/L)
0	5.50
41	4.91
48	4.81
55	4.69
75	4.38
96	4.12
127	3.84
146	3.62
162	3.59
180	3.44
194	3.34
212	3.27
267	3.07
318	2.92
379	2.84
410	2.79
∞	2.60

Equations 6.18 and 6.19 can be combined to give

$$\frac{dc_A}{dt} = k_1 \left[\frac{4}{K_{eq}} (c_{A0} - c_A)^2 - c_A^2 \right], \tag{6.20}$$

or, formally separating the concentration- and time-dependent terms,

$$\frac{K_{eq} dc_A}{4(c_{A0} - c_A)^2 - K_{eq} c_A^2} = k_1 dt.$$

The left-hand side is a form that is readily found in tables of integrals. Upon integration of the left-hand side from c_{A0} to the current value $c_A(t)$, and the right-hand side from $t = 0$ to the present time, we obtain

$$\ln \left[\frac{c_A(2 - \sqrt{K_{eq}}) - 2c_{A0}}{c_A(-2 - \sqrt{K_{eq}}) + 2c_{A0}} \right] = \frac{4c_{A0}k_1}{\sqrt{K_{eq}}} t. \tag{6.21}$$

EXAMPLE 6.4 The reaction between sulfuric acid and diethyl sulfate studied in Example 6.1 is, in fact, reversible, although the assumption of irreversibility gave a good fit to the data up to a time of 194 minutes. The full data set is shown in Table 6.5. $c_{A0} = c_{B0}$. Find the rate expression, assuming that both the forward and reverse reactions may be described by mass action kinetics.

From the data given in Table 6.5, $K_{eq} = 4 \left(\frac{5.50}{2.60} - 1 \right)^2 \approx 5$. The data in Table 6.5 are plotted according to Equation 6.21 in Figure 6.6, with $c_{A0} = 5.50$ and $K_{eq} = 5$. The data do follow a straight line and are consistent with the

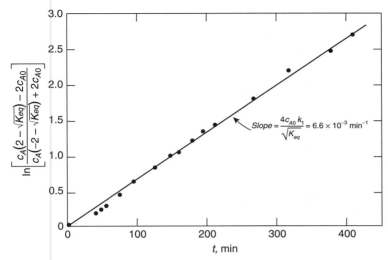

Figure 6.6. Computation of the forward rate constant for the reaction between sulfuric acid and diethyl sulfate in aqueous solution.

assumption of mass action kinetics. Calculation of k_1 from the slope yields $k_1 = 6.7 \times 10^{-4}$ L/(g-mol min). This value differs by only 10 percent from the value obtained in Example 6.1 by assuming irreversibility over the first 194 minutes.

6.8 Concluding Remarks

The important concept here is the rate of reaction, which addresses the fact that the mass of a component species is not conserved in the balance equations. The particular rate constitutive equations used in this chapter are quite elementary; mass action kinetics may be followed in systems of interest, but the kinetics may also be far more complex because of chemical steps that are not obvious from the overall stoichiometry. The basic principles developed here are sufficient to enable us to address meaningful design problems in the next chapter despite the elementary forms of the rate equations. More complex reactions will typically be considered in a subsequent course in the chemical engineering core.

Bibliographical Notes

Basic textbooks on physical chemistry contain at least one chapter on the rates of chemical reactions. The classic textbook treatments of reaction kinetics are by Laidler and by Frost and Pearson, the latter now in a third edition as Moore and Pearson:

Laidler, K. J., *Chemical Kinetics*, 3rd Ed, Prentice-Hall, Englewood Cliffs, NJ, 1987.

Moore, J. W., and R. G. Pearson, *Kinetics and Mechanism*, 3rd Ed., Wiley, New York, 1981.

More complex reaction schemes and reactor geometries than those covered in this chapter are an essential part of the chemical engineering core course that covers the subjects of kinetics and reactor design.

PROBLEMS

6.1. Suppose that the data in Table 6.1 for the reaction of sulfuric acid with diethyl sulfate in aqueous solution are described by a rate of the form $r = kc_A^n$. Estimate n using the methodology of Appendix 2A. Investigate the consequences of an error of 1 percent in determining c_A.

6.2. Data for the decomposition of dibromosuccinic acid (2,3-dibromo-butanedioic acid), $C_2H_2Br_2(COOH)_2$, in a batch reactor are shown in Table 6.P1 as mass of acid remaining as a function of time. Find the rate expression.

Table 6.P1.

Time (min)	Mass of acid (g)
0	5.11
10	3.77
20	2.74
30	2.02
40	1.48
50	1.08

6.3. Reconsider the data in Table 6.1 for the reaction between sulfuric acid and diethyl sulfate, and suppose that you believe that the forward reaction rate is actually of the form $r = k'c_A$. Test this assumption. What conclusion do you draw?

6.4. The hydrolysis of acetic anhydride in excess water to form acetic acid, $(CH_3CO)_2O + H_2O \rightarrow 2CH_3COOH$, was studied by Elridge and Piret, who found that the rate at $15°C$ is first order in anhydride, with $k = 0.0567 \text{ min}^{-1}$. A batch reactor initially holds 100 kg of anhydride. How long will it take for 90 kg to be converted to acid? 99 kg?

6.5. The reaction $ClO_3^- + 3H_2SO_4 \rightarrow Cl^- + 3SO_4^= + 6H^+$ in 0.2N H_2SO_4 was studied in 1932 by Nixon and Krauskopf, who reported the data in Table 6.P2. Find a rate expression that is consistent with the data. (Hint: Does $r = kc_A c_B$ work?)

Table 6.P2.

t (min)	ClO_3^- (mol/L)	H_2SO_4 (mol/L)
0	0.0160	0.0131
2	0.0146	0.0090
3.5	0.0141	0.0074
5	0.0136	0.0060
7.5	0.0131	0.0044
10	0.0127	0.0034
12.5	0.0125	0.0026
15	0.0123	0.0020
17.5	0.0121	0.0015

6.6. Show that Equation 6.11a for the reaction A + B → μM with equal initial concentrations of A and B ($\psi = 0$) follows from Equation 6.11b for unequal initial concentrations in the limit $\psi \to 0$.

6.7. The data in Table 6.P3 are for reactant A in the irreversible reaction A + βB → products. The initial concentrations are believed to be in stoichiometric balance. Estimate the reaction order and find the rate constant.

Table 6.P3.

time (min)	c_A (mol/L)
0	10
3	7.31
8	4.71
10	3.54
12	3.17
14	2.40
19	1.25
22	0.98
25	0.89
35	0.21

6.8. The reaction acetic acid (A) + butanol (B) → ester + water was studied by Leyes and Othmer, who found the data in Table 6.P4 at 100°C in the presence of 0.03 percent sulfuric acid. The initial concentrations were 0.2327 mol acetic acid/100 g of solution and 1.583 mol butanol/100 g of solution. Leyes and Othmer claimed that the reaction is second order in acetic acid and zero order in butanol up to 75 percent conversion; i.e., $r = kc_A^2$. Do the data support this contention? Can you distinguish this relation from $r = kc_A c_B$ with the data given here? (You may assume that the density of the solution is constant over the course of the reaction with a value of approximately 750 kg/m^3. Why is this assumption permissible?)

Table 6.P4.

time (hr)	moles of acid converted/100 g solution
0	0
1	0.1552
2	0.1876
3	0.2012
4	0.2067
5	0.2089
6	0.2099
7	0.2109

6.9. Consider the full data set in Table 6.5 for the reversible reaction between sulfuric acid and diethyl sulfate and test the applicability of each of the following rate expressions:

a. $r_{A-} = k_1 c_A c_B,\quad r_{A+} = k_2 c_M$
b. $r_{A-} = k_1 c_A,\quad r_{A+} = k_2 c_M$
c. $r_{A-} = k_1 c_A,\quad r_{A+} = k_2 c_M^2.$

Comment on the results.

6.10. The saponification of the ester propargyl acetate, $(A, H_2C{=}\overset{\displaystyle H}{\underset{\displaystyle |}{C}}{-}\overset{\displaystyle O}{\underset{\displaystyle \|}{C}}{-}O{-}C{\equiv}CH).$ with base (B, OH$^-$) was studied by Myers, Collett, and Lazzell using a conductivity technique to follow the course of the reaction. Conductivity, $C(t)$, is related to the conversion of the base through the relation

$$\text{Fraction base converted} \ = \ \frac{c_B(t)}{c_{B0}} = \frac{C(t) - C(\infty)}{C(0) - C(\infty)}.$$

They report the data in Table 6.P5 for a solution that is initially $0.00873\ N$ ester and $0.00673\ N$ base. Find a rate expression that is consistent with these data.

Table 6.P5.

t (min)	C
0	1.087
0.200	1.042
0.417	1.000
0.737	0.952
1.047	0.909
1.397	0.870
1.787	0.833
2.187	0.800
2.637	0.769
3.147	0.741
3.667	0.714
∞	0.490

6.11. The reaction A $\rightarrow$ products is believed to follow a rate expression $r_{A-} = \frac{k_1 c_A}{1 + k_2 c_A}$. How will the concentration in a batch reactor change with time? Discuss the limits $k_2 c_{A0} \gg 1$ and $k_2 c_{A0} \ll 1$.

6.12. The formation of sugars from biomass is of considerable interest in a variety of applications, including energy alternatives to fossil fuel. Harris and Kline studied the formation of glucose from cellulose obtained from Douglas Fir in 1949, carrying out the reaction at a number of temperatures in the presence of HCl.

  a. The rate was found to be first order in cellulose (A) concentration ($r = k_1 c_A$), but dependent on HCl concentration as shown in Table 6.P6. Find the dependence of the rate on HCl. (Hint: try $k_1 = k_1' c_{HCl}{}^n$.)
  b. Glucose also decomposes in the presence of HCl. Harris and Kline reported the decomposition reaction to be first order in glucose ($r_{glucose-} = k_2 c_{glucose}$), with a first-order rate constant at 190°C that depended on HCl as shown in Table 6.P7. Can you find a rate expression in a simple form? (Hint: Consider the form in Problem 6.11.)

Table 6.P6.

HC1 concentration (mol/L)	k_1 (min^{-1})	
	160°C	190°C
0.055	0.00203	0.0627
0.11	0.00486	0.149
0.22	0.01075	0.357
0.44	0.0261	
0.88	0.0672	

Table 6.P7.

HC1 concentration (mol/L)	k_2 (min^{-1})
0.055	0.0488
0.11	0.107
0.22	0.218
0.44	0.406
0.88	0.715

Appendix 6A: Constant Density Assumption

As in Chapter 4, we carry out the analysis of the reactor overall mass balance for the case in which the density is a linear function of the concentration. We will do the calculation for the special case represented by Equations 6.2 and 6.4, but the result is completely general. We assume that the species of interest are dissolved in a nonreactive solvent, and that we can write the density of the mixture as

$$\rho = \rho_0 + \phi_A c_A + \phi_B c_B + \phi_M c_M. \tag{6A.1}$$

Equation 6.1 then becomes

$$\rho_0 \frac{dV}{dt} + \phi_A \frac{dc_A V}{dt} + \phi_B \frac{dc_B V}{dt} + \phi_M \frac{dc_M V}{dt} = \rho_0(q_{Af} + q_{Bf} - q) + \phi_A(c_{Af}q_{Af} - c_A q)$$

$$+ \phi_B(c_{Bf}q_{Bf} - c_B q) + \phi_M(-c_A q). \tag{6A.2}$$

Multiplying Equations 6.3a, b, and c by ϕ_A, ϕ_B, and ϕ_C, respectively, and combining with Equation 6A.2 gives

$$\frac{dV}{dt} = q_{Af} + q_{Bf} - q + \frac{Vr}{\rho_0}(\phi_A + \beta\phi_B - \mu\phi_M). \tag{6A.3}$$

The last term on the right of Equation 6A.3 represents the contribution to volume change from the chemical reaction. The rate of change of density with composition is usually nearly proportional to the molecular weight of the dissolved species. Hence, we can write

$$\phi_A = \phi M_{wA}, \quad \phi_B = \phi M_{wB}, \quad \phi_M = \phi M_{wM}, \qquad (6A.4)$$

where M_{wi} denotes the molecular weight of species i. Equation 6.A3 becomes

$$\frac{dV}{dt} = q_{Af} + q_{Bf} - q + \frac{Vr\phi}{\rho_0}(M_{wA} + \beta M_{wB} - \mu M_{wM}). \qquad (6A.5)$$

But from the reaction stoichiometery, $M_{wA} + \beta M_{wB} = \mu M_{wM}$. Thus,

$$\frac{dV}{dt} = q_{Af} + q_{Bf} - q, \qquad (6A.6)$$

which is the result that would be obtained for a system with a density that is independent of concentration. Note that this result, which will usually be a reasonable approximation for liquid systems, depends on *two* assumptions:

1. Density is linear in concentration (Equation 6A.1).
2. Rate of change of density with concentration of each species is proportional to the molecular weight of that species (Equation 6A.4).

7 Designing Reactors

7.1 Design

Engineering is ultimately about making things for the benefit of society. We typically use the term *design* for the set of instructions by which a craftsperson can turn an idea into an object, and that meaning of the word carries over to the engineering activity of making tangible objects. Design has an additional meaning for a chemical engineer, however; it is the conceptualization of a *process* for manufacturing something – a chemical, perhaps. In this use of the word we address the problem of designing what pieces of equipment are needed, how they should be connected, how large they should be, and so on, but we do not address the question of how the equipment should be manufactured – materials of construction, locations of welds, precise geometry, and so forth. It is the second sense of the word that we employ in this chapter. Thus, we will exploit the understanding of reaction kinetics developed in Chapter 6 to determine reactor sizes and material flow rates, but we will view the reactor simply as a vessel of a given size, with no attention to the detail that would be required for actual fabrication.

Process design is a subject that is traditionally taught as a *capstone* course during the last year of a chemical engineering program, because a complete approach to process design obviously requires a broad base of understanding of chemical engineering fundamentals. There is no reason to put the elements of design off, however, and we will address some typical, albeit simplified, problems in this chapter. Our goal is to explore one of the basic issues in design, namely that the conservation and constitutive equations must be tied in to other considerations – economic, environmental, and so on – in order to arrive at an engineering solution to the problem of making things. We will do this through reacting systems that are converted in continuous-flow stirred-tank reactors (CFSTRs). We start slowly, with some elementary reactor calculations that are an essential underpinning to the more interesting engineering issues that we can then attack.

7.2 The CFSTR

For illustrative purposes we will consider the single irreversible reaction $A + B \rightarrow \mu M$ in the liquid phase in CFSTRs, where the reactants A and B are brought into the reactor in separate streams. We have already derived the governing equations for this reaction system (with slightly more general stoichiometry) in Chapter 6, with the following results:

$$\frac{dV}{dt} = q_{Af} + q_{Bf} - q, \tag{7.1}$$

$$\frac{d}{dt}c_A V = c_{Af}q_{Af} - c_A q - rV, \tag{7.2a}$$

$$\frac{d}{dt}c_B V = c_{Bf}q_{Af} - c_B q - rV, \tag{7.2b}$$

$$\frac{d}{dt}c_M V = -c_M q + \mu rV. \tag{7.2c}$$

For definiteness, when required, we will assume that the kinetics are mass action, with $r = kc_A c_B$.

For design purposes we assume that the system operates in the steady state, so all time derivatives are equal to zero. The operating equations are therefore

$$q = q_{Af} + q_{Bf}, \tag{7.3}$$

$$0 = c_{Af}q_{Af} - c_A(q_{Af} + q_{Bf}) - rV, \tag{7.4a}$$

$$0 = c_{Bf}q_{Af} - c_B(q_{Af} + q_{Bf}) - rV, \tag{7.4b}$$

$$0 = -c_M(q_{Af} + q_{Bf}) + \mu rV. \tag{7.4c}$$

where we have made use of Equation 7.3 in writing Equations 7.4a,b, and c. Note that, taking the parameters of the reaction rate as known, we have three equations for the eight variables $c_{Af}, c_{Bf}, q_{Af}, q_{Bf}, V, c_A, c_B,$ and c_M. Thus, five quantities must be specified, with the other three determined from Equations 7.4a,b, and c.

We can simplify these equations for many relevant calculations. We define two quantities:

$$\theta = V/q = V/(q_{Af} + q_{Bf}), \tag{7.5a}$$

$$\lambda = q_{Af}/q = q_{Af}/(q_{Af} + q_{Bf}). \tag{7.5b}$$

θ is the *residence time*, the mean time that a fluid element spends in a perfectly mixed tank, which we have seen before. λ is simply the ratio of the volumetric flow rate of the feedstream of A to the total volumetric flow. We can now write Equations 7.4a,b, and c as

$$0 = \lambda c_{Af} - c_A - r\theta, \tag{7.6a}$$

$$0 = (1 - \lambda)c_{Bf} - c_B - r\theta, \tag{7.6b}$$

$$0 = -c_M + \mu r\theta. \tag{7.6c}$$

Equations 7.6a and 7.6b can be combined to give

$$c_A = c_B + \lambda c_{Af} - (1 - \lambda)c_{Bf}, \tag{7.7a}$$

whereas Equations 7.6b and 7.6c can be combined to give

$$c_M = -\mu c_B + (1 - \lambda)\mu c_{Bf}. \tag{7.7b}$$

We have now reduced the number of variables by one, and we have three independent equations, so four quantities must be specified from among the seven: $c_{Af}, c_{Bf}, \lambda, \theta, c_A, c_B,$ and c_M.

Note that the reaction rate does not appear in Equations 7.7a and b, so these equations simply reflect the reaction stoichiometry which, together with the feed distributions, completely establishes the relations between the three compositions; that is, for given feed conditions, only one composition is independent. Note also that, because the reaction is irreversible, Equation 7.6c (or Equation 7.7b) is uncoupled from the rest of the equations if c_M is one of the variables that is sought as part of the solution. Finally, it is clear from Equation 7.7b that the four variables $\lambda, c_{Bf}, c_B,$ and c_M cannot all be specified independently, so the choice of the four variables that we can fix is not arbitrary.

Now, suppose we wish to design a reactor for this reaction. It seems obvious that we would want to have stoichiometric conditions ($c_A = c_B$) in the reactor, since the reaction is irreversible. (Why is irreversibility relevant here? What might be different if the reaction were reversible?) It then follows from Equation 7.7a that $\lambda c_{Af} = (1 - \lambda)c_{Bf}$. For definiteness we will take the feed flow rates to be equal ($\lambda = \frac{1}{2}$), in which case c_{Af} must equal c_{Bf}, and we take the feed conditions to be specified. We can specify only one more quantity, and the choice depends on our objective. If we must meet a given production schedule of product, then qc_M is fixed. If the reactor already exists, as will often be the case, then V is fixed. We immediately see an element of ambiguity here: q and V appear in the equations only in the ratio $\theta = V/q = V/(q_{Af} + q_{Bf})$, but we will need to consider them separately if we wish to address actual production issues.

EXAMPLE 7.1 Suppose that we have a reaction with mass action kinetics, where the rate constant $k = 6.05 \times 10^{-4}$ L/(g-mol min), as found for the reaction between sulfuric acid and diethyl sulfate in Example 6.1. Assume that the feed concentrations are specified, with $c_{Af} = c_{Bf} = 11$ g-mol/L $= 110$ kg-mol/m^3. How does the conversion depend on the residence time θ?

We replace r with the mass action expression kc_A^2 ($c_A = c_B$ because of balanced stoichiometry) in Equation 7.6a and solve the quadratic equation for c_A, after which we obtain c_M from Equation 7.7b. The resulting conversions are shown in Table 7.1 for a range of values of θ, where the residence time is given in minutes. Clearly, the larger the residence time, the higher the conversion of the feed and the higher the concentration of product. (The reactants spend more time on average in contact in the reactor, so they have a greater opportunity to react.) The relation between conversion and residence time is nonlinear.

Table 7.1. *Product concentrations for various residence times.*

θ (min)	c_A (kg-mol/m^3)	c_M (kg-mol/m^3)
127	41.5	27
84.7	44.5	21
63.5	47	16
50.9	48	14
42.8	49	12
31.8	50	10
25.4	50.5	9

For the reaction under consideration here, a substantial amount of unreacted feed – from 38 to 46 percent – would have to be recovered from the product stream over the entire range of residence times considered, so the cost of constructing and operating the separation system would be an important economic factor.

EXAMPLE 7.2 Now, for definiteness, we assume that we have a reactor of fixed volume, which we take to be 762 L = 0.762 m^3. Find the productivity qc_M (the rate at which M is produced as effluent from the reactor) as a function of flow rate.

Since the reactor volume is specified, we compute the required flow rate from $q = V/\theta$ and add these two additional terms to Table 7.1. The results are shown in Table 7.2.

We find in Example 7.2 that the productivity is highest with the highest throughput and the lowest conversion of the reactants, hence with the highest separation costs! On the other hand, if we were to fix the throughput and calculate the required reactor size, then the highest productivity would occur for the largest reactor; for the data used to construct Table 7.1 and $q = 6 \times 10^{-3}$ m^3/min, for example, the maximum productivity would be $qc_M = 0.162$ kg-mol/min for $\theta = 127$ and $V = 762$ L. The point of this exercise is that the design depends on the question that is asked, and the result can be very different, depending on what is fixed in the system.

Table 7.2. *Results for conditions in Table 7.1, but with $V = 762$ L = 0.762 m^3.*

q(m^3/min)	θ (min)	c_A (kg-mol/m^3)	c_M (kg-mol/m^3)	qc_M (kg-mol/min)
6×10^{-3}	127	41.5	27	0.162
9×10^{-3}	84.7	44.5	21	0.189
12×10^{-3}	63.5	47	16	0.192
15×10^{-3}	50.9	48	14	0.210
18×10^{-3}	42.8	49	12	0.216
24×10^{-3}	31.8	50	10	0.240
30×10^{-3}	25.4	50.5	9	0.270

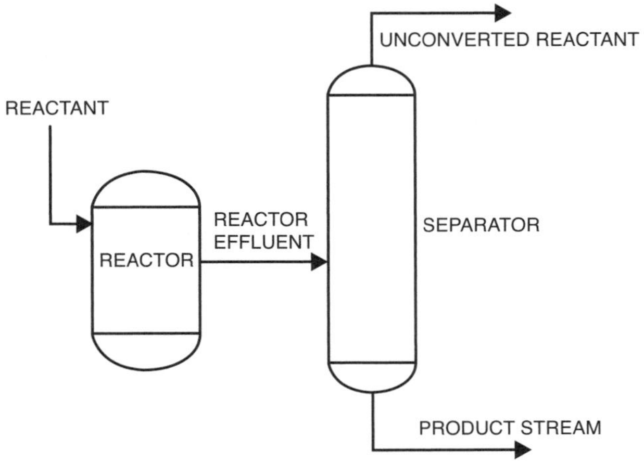

Figure 7.1. Schematic of a reactor-separator system. The unconverted reactant stream is assumed to be sufficiently pure to be used as part of the feed to the reactor.

7.3 An Optimal Design Problem

We are now in a position to consider a realistic, if simplified, problem in optimal design for a reacting system. The process is shown in Figure 7.1. A reaction is carried out in a stirred-tank reactor. The effluent stream is then taken to a separator. We need not be specific here about the type of separation that is to be used, which might be distillation, crystallization, membrane processing, and so on, but we will assume that the separation is very efficient and leads to relatively pure streams of A and M. We therefore assume that the unreacted A can be recovered and reused in the reactor feed. For simplicity, in order to obtain equations that can be manipulated analytically and thus provide transparent solutions that reveal the important features of the approach, we will consider the idealized reaction system A $\rightarrow$ M, with first-order irreversible kinetics, $r = kc_A$. We suppose that the required production rate of M is specified, perhaps based on market projections; hence, qc_M is a fixed quantity, which we denote p (for *production rate*).

The trade-offs in the design problem are fairly clear. The larger the reactor, the higher the conversion will be for a given throughput rate. Hence, all other things being equal, the largest possible reactor will result in the smallest possible throughput rate, hence the smallest amount of material to be handled by the downstream separation system. The cost of operation depends on throughput, so low throughput means low operating cost. But the larger the reactor, the greater the capital cost. The trade-off in the design is therefore between the cost of construction and the cost of operation. We will make this trade-off quantitative in what follows.

The CFSTR equations for the first-order reaction A $\rightarrow$ M follow directly from Equations 7.1 and 7.2 by setting $q_{Bf} = 0$. At steady state we therefore obtain

$$0 = qc_{Af} - qc_A - kc_AV, \tag{7.8a}$$

$$0 = -qc_M + kc_AV. \tag{7.8b}$$

It is useful to rearrange these equations, making use of the fact that the production rate $p \equiv qc_M$ is a fixed quantity. From Equation 7.8b we obtain

$$V = \frac{p}{kc_A} > \frac{p}{kc_{Af}} \equiv V_{\min}. \qquad (7.9a)$$

The inequality follows from the fact that c_A must always be smaller than the feed concentration c_{Af}. Hence, for a given feed concentration and a fixed production rate, there is a lower bound to the reactor size that is necessary in order to carry out the conversion. Similarly, by adding Equations 7.8a and 7.8b we eliminate the reaction rate term and obtain a stoichiometric equation that can be rearranged to the form

$$q = \frac{p}{c_{Af} - c_A} > \frac{p}{c_{Af}} \equiv q_{\min}. \qquad (7.9b)$$

Hence, there is a minimum throughput rate in order to achieve the required production rate. Clearly, operating with a flow rate close to $q_{\min}$ would require a reactor with a very large volume in order to achieve nearly complete conversion of the reactant A, whereas operating with a volume close to $V_{\min}$ would require a very large flow rate in order to produce the required product despite the small conversion.

Let us now formulate the optimal design problem. The overall return, which we will denote $\mathfrak{R}$, is given by the income from the sale of the product less the cost of raw materials, the capital cost of constructing the equipment, and the cost of operation. We can calculate each of these quantities on an annual basis over the projected lifetime of the process, or we can calculate the net present value of each as described in Section 3.2. The approach to the calculation is the same in either case.

The return from the sale of product is fixed, because the production rate is fixed and we assume that everything will be sold at a known price. Hence, this term will simply enter the return function as a constant. The capital cost of the reactor will depend on the volume. For simplicity, we will assume that the cost is *proportional* to the volume; this is not a good assumption, because costs generally do not increase linearly with volume, but it will suffice for illustrative purposes. Using a better cost function would only change the algebra. Hence, we write

$$\text{Capital cost of reactor} = C_V V = \frac{C_V p}{kc_A}.$$

The size of the separation unit, hence the cost, will depend on the throughput; we will take the cost as proportional to throughput rate and write

$$\text{Capital cost of separator} = C_D q = \frac{C_D p}{c_{Af} - c_A}.$$

The operating costs for both units will have some fixed amount that is independent of operating conditions, and this can be included in the fixed term in the return function; the variable costs will depend on the amount of material processed, and we will take this term to be proportional to q. We therefore write

$$\text{Variable costs of operation} = C_O q = \frac{C_O p}{c_{Af} - c_A}.$$

Finally, the cost of raw material will be proportional to the amount of A that is used. We assume here that the separation is very efficient and that unreacted A can be reused. The net cost is therefore proportional only to the reactant that is used to produce product; that is, the cost is proportional to $q(c_{Af} - c_A) = qc_M = p$, so with this assumption the cost of reactant is fixed and can be combined with the fixed term in the return function. We thus obtain

$$\mathfrak{R} = \text{fixed terms} - \frac{C_D p}{c_{Af} - c_A} - \frac{C_O p}{c_{Af} - c_A} - \frac{C_V p}{k c_A}$$

$$= \text{fixed terms} - \frac{M_{TC}}{c_{Af} - c_A} - \frac{M_{VC}}{c_A}. \tag{7.10}$$

The terms included in M_{TC} are associated with throughput costs, while M_{VC} contains the reactor volume costs.

The optimization problem thus reduces to the selection of the conversion in the reactor. Once c_A is fixed, the reactor volume V and the throughput q are determined. The optimal value of c_A is clearly somewhere between the two extremes of c_{Af} and zero; in the former case the throughput costs become infinite, whereas in the latter case the reactor cost becomes infinite. Hence, the function $\mathfrak{R}$ must have a maximum for a value of c_A somewhere between zero and c_{Af}. We can find the maximum in $\mathfrak{R}$ by setting the derivative with respect to c_A to zero. (How do we know that this will give us a maximum, and not a minimum?) We thus write

$$\frac{d\mathfrak{R}}{dc_A} = -\frac{M_{TC}}{(c_{Af} - c_A)^2} + \frac{M_{VC}}{c_A^2} = 0. \tag{7.11}$$

It is convenient to define the fraction of unreacted A as $x = c_A/c_{Af}$. Equation 7.11 can then be written

$$\frac{(1-x)^2}{M_{TC}} = \frac{x^2}{M_{VC}}, \tag{7.12}$$

with the solution

$$x = \frac{1}{1 + \left(\dfrac{M_{TC}}{M_{VC}}\right)^{1/2}}. \tag{7.13}$$

(Both square roots must have the same sign to ensure $0 \leq x \leq 1$.)

The unreacted fraction of A for the optimal design is shown as a function of the relative cost terms in Table 7.3, together with the fractional conversion $(1 - x)$, the volume relative to the minimum volume, the throughput rate relative to the minimum throughput rate, and the residence time $(\theta = V/q)$ multiplied by the first-order reaction rate constant. The last three design variables (only two of which are independent) are computed as follows:

$$\frac{V}{V_{\min}} = \frac{1}{x}, \quad \frac{q}{q_{\min}} = \frac{1}{1-x}, \quad k\theta = \frac{1-x}{x} = \left(\frac{M_{TC}}{M_{VC}}\right)^{1/2}. \tag{7.14}$$

(The values of M_{TC}/M_{VC} are not evenly spaced in the table because it is more instructive to have evenly spaced conversions.)

Table 7.3. *Optimal reactor design parameters as functions of the relative cost parameters M_{TC}/M_{VC}.*

$\dfrac{M_{TC}}{M_{VC}}$	Fractional conversion		$\dfrac{V}{V_{\min}}$	$\dfrac{q}{q_{\min}}$	$k\theta$
	$x = c_A/c_{Af}$	$1-x$			
0	1.0	0	1.0	∞	0
0.012	0.90	0.10	1.11	10.00	0.11
0.063	0.80	0.20	1.25	5.00	0.25
0.184	0.70	0.30	1.43	3.33	0.43
0.250	0.67	0.33	1.50	3.00	0.50
0.290	0.65	0.35	1.54	2.86	0.54
0.444	0.60	0.40	1.67	2.50	0.67
0.669	0.55	0.45	1.82	2.22	0.82
1.0	0.50	0.50	2.00	2.00	1.0
1.49	0.45	0.55	2.22	1.82	1.22
2.25	0.40	0.60	2.50	1.67	1.50
3.45	0.35	0.65	2.86	1.54	1.86
4.00	0.33	0.67	3.00	1.50	2.00
5.44	0.30	0.70	3.33	1.43	2.33
16.0	0.20	0.80	5.00	1.25	4.00
81.0	0.10	0.90	10.00	1.11	9.00
∞	0	1.0	∞	1.0	∞

The results in Table 7.3 are very revealing. When the parameters for costs associated with throughput and reactor volume are equal, the optimum is to operate with 50 percent conversion, with a volume equal to twice the minimum volume and a throughput rate equal to twice the minimum throughput, resulting in a reactor residence time equal to the reciprocal of the first-order rate constant. The volume and throughput differ by no more than 50 percent from the respective minimum values as long as the relative costs are within a factor of four of one another. These are not especially tight bounds, but they set reasonable limits on the likely design range even before we have precise values for the cost factors. Note that we have established the optimal value for the return:

$$\mathfrak{R} = \text{fixed terms} - \frac{p}{c_{Af}}\left(\frac{C_D + C_O}{1-x} + \frac{C_V}{kx}\right), \tag{7.15}$$

where x is given in terms of the cost parameters by Equation 7.13. The return will only be positive, of course, and the process financially feasible, if the term on the right is less than the net of the fixed terms.

This example of an optimal design problem is quite simplified, as we have attempted to make clear throughout: The reaction is much simpler than any that is likely to be of interest for a real design; the cost equations, which are taken to be linear functions of the relevant variables, are grossly oversimplified; the separation is assumed to be perfect; and we have not taken into account the fact that temperature affects the rate constant and would be included as an additional design variable in a realistic situation, thus introducing additional cost trade-offs (faster rate and

Table 7.4. *Typical reactions following the scheme of Equation 7.16.*

Reactants		Products		
A	B	R	S	T
Water	Ethylene oxide*	Ethylene glycol	Diethylene glycol	Triethylene glycol
Ammonia	Ethylene oxide*	Monoethanolamine	Diethanolamine	Triethanolamine
Methyl, ethyl, or butyl alcohol	Ethylene oxide*	Monoglycol ether	Diglycol ether	Triglycol ether
Benzene	Chlorine	Monochloro-benzene	Dichlorobenzene	Trichlorobenzene
Methane	Chlorine	Methyl chloride	Dichloromethane	Trichloromethane

* Also carried out using propylene oxide.

smaller reactor vs. increased energy cost, for example). Nonetheless, the example includes the essential features of a typical process design, and it correctly shows many of the issues that will be faced in more realistic situations. First, we see that the conversion in the reactor is the primary factor affecting the overall design, including "downstream" processing, which is very typical. We see the critical role played by the relative costs for volume and throughput. We see that we can get rough estimates by making rather idealized assumptions, such as perfect separation and flow rates and volumes that are roughly twice the minimum possible values. (If the process is not feasible with assumptions like these – i.e., if the return $\Re$ is negative – then the process is unlikely to be feasible with less idealized assumptions.)

7.4 Product Selectivity

Most reacting systems of interest involve multiple reactions, and *product selectivity* – that is, the distribution of the various products – is a primary factor in the design. In this section we will consider a very realistic case. The following sequence of reactions, all of which proceed nearly irreversibly with mass-action kinetics, accounts for the production of more than 3×10^9 kg/year (3 million metric tons) in the United States alone, and more than 9×10^9 kg/year in the world:

$$A + B \to R, \quad R + B \to S, \quad S + B \to T. \tag{7.16}$$

Some typical reactants and products within this class are listed in Table 7.4.

We will focus here on the reaction between water and ethylene oxide (*EtO*, also known by the official IUPAC name *oxirane*) to form mono-, di-, and triethylene glycol, as shown in Figure 7.2. The glycols are colorless, odorless liquids at room temperature. Sixty percent of the ethylene oxide used in the United States is consumed as a reactant in this reaction scheme. The monoglycol, usually called ethylene glycol (and sometimes MEG), is the primary product, but the di- and triglycols also have industrial uses. About two-thirds of the ethylene glycol manufactured worldwide is used as a chemical intermediate in the manufacture of polyester resins for fibers,

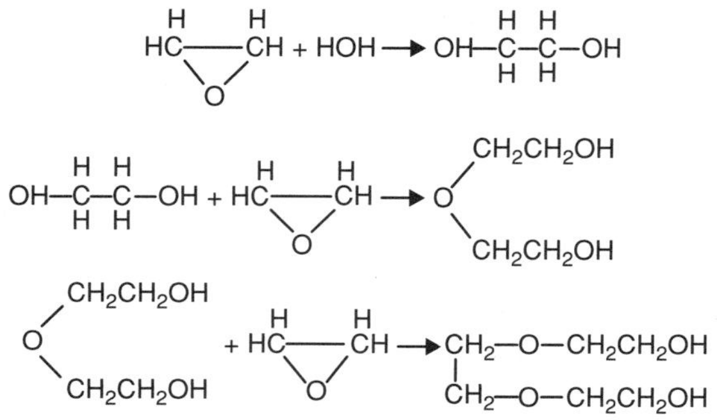

Figure 7.2. Reaction scheme for the reaction of ethylene oxide and water to form ethylene glycols.

films, and bottles, while about one-fourth is used as antifreeze in engine coolants. The distribution of products by weight is given by various sources as being in the neighborhood of 88/10/2 mono/di/tri and 90/9/1; as we shall see, the product distribution is determined by the reactor design. In one year for which specific figures are available, the recorded market demand in the United States was 90/8/2.

We assume that the reactants A (water) and B (ethylene oxide) are mixed prior to the reactor and enter in a single feedstream, and we assume that no product is present in the feedstream. With the usual assumptions regarding the density we then obtain the steady-state equations for a CFSTR:

$$\text{A (water): } 0 = q(c_{Af} - c_A) - r_{A-}V, \tag{7.17a}$$

$$\text{B (EtO): } 0 = q(c_{Bf} - c_B) - r_{B-}V, \tag{7.17b}$$

$$\text{R (MEG): } 0 = -qc_R + r_{R+}V - r_{R-}V, \tag{7.17c}$$

$$\text{S (di-glycol): } 0 = -qc_S + r_{S+}V - r_{S-}V, \tag{7.17d}$$

$$\text{T (tri-glycol): } 0 = -qc_T + r_{T+}V. \tag{7.17e}$$

With mass action kinetics, which have been validated for this reaction system, the rates are as follows:

$$r_{A-} = r_{R+} = k_1 c_A c_B, \; r_{R-} = r_{S+} = k_2 c_R c_B, \; r_{S-} = r_{T+} = k_3 c_S c_B,$$
$$r_{B-} = r_{A-} + r_{R-} + r_{S-}.$$

The reactor equations are then

$$\text{A: } 0 = q(c_{Af} - c_A) - k_1 c_A c_B V, \tag{7.18a}$$

$$\text{B: } 0 = q(c_{Bf} - c_B) - c_B(k_1 c_A + k_2 c_R + k_3 c_S)V, \tag{7.18b}$$

$$\text{R: } 0 = -qc_R + c_B(k_1 c_A - k_2 c_R)V, \tag{7.18c}$$

$$\text{S: } 0 = -qc_S + c_B(k_2 c_R - k_3 c_S)V, \tag{7.18d}$$

$$\text{T: } 0 = -qc_T + k_3 c_S c_B V. \tag{7.18e}$$

It is convenient to express the product concentrations in terms of $x_A = c_A/c_{Af}$, which is the fraction of unreacted A; that is, one minus the fractional conversion of water.

For example, we can solve Equation 7.18c for c_B and substitute directly into Equation 7.18a to obtain, after some algebraic manipulation,

$$\frac{c_R}{c_{Af}} = \frac{x_A(1 - x_A)}{x_A + \frac{k_2}{k_1}(1 - x_A)}. \qquad (7.19a)$$

Similarly, solving for c_B in Equations 7.18d and 7.18e, respectively, and substituting into Equation 7.17a,

$$\frac{c_S}{c_{Af}} = \frac{\frac{k_2}{k_1}x_A(1 - x_A)^2}{\left[x_A + \frac{k_2}{k_1}(1 - x_A)\right]\left[x_A + \frac{k_3}{k_1}(1 - x_A)\right]}, \qquad (7.19b)$$

$$\frac{c_T}{c_{Af}} = \frac{\frac{k_2 k_3}{k_1^2}(1 - x_A)^3}{\left[x_A + \frac{k_2}{k_1}(1 - x_A)\right]\left[x_A + \frac{k_3}{k_1}(1 - x_A)\right]}. \qquad (7.19c)$$

By similar manipulations we obtain the expression for the conversion of B:

$$\frac{c_{Bf} - c_B}{c_{Af}} = \frac{x_A(1 - x_A)\left[x_A + \frac{k_3}{k_1}(1 - x_A)\right] + 2\frac{k_2}{k_1}x_A(1 - x_A)^2 + 3\frac{k_2 k_3}{k_1^2}(1 - x_A)^3}{\left[x_A + \frac{k_2}{k_1}(1 - x_A)\right]\left[x_A + \frac{k_3}{k_1}(1 - x_A)\right]}. \qquad (7.19d)$$

Note that these relations depend only on the conversion of A and the relative rate constants but are independent of the reactor volume and throughput. They are not stoichiometric relations, because the rates are explicitly included, but they lead to the following powerful conclusion: *Any reactor conditions in a CFSTR that produce a given conversion of water will have the same product distribution.*

The rate constant k_1 for this system has a value of 6.37×10^{-7} L/(g-mol min) at 25°C, and, to within experimental uncertainty, the ratios of the rate constants are $k_2/k_1 = k_3/k_1 = 2.0$. (We need only the latter result to evaluate the product distribution in terms of x_A.) The distributions of the products by mass fraction are given in Table 7.5 in terms of x_A, where the mass fraction of species i is calculated as follows:

$$\text{Mass fraction of } i = \frac{\frac{c_i}{c_{Af}}M_{wi}}{\frac{c_R}{c_{Af}}M_{wR} + \frac{c_S}{c_{Af}}M_{wS} + \frac{c_T}{c_{Af}}M_{wT}}.$$

The product molecular weights are $M_{wR} = 62$, $M_{wS} = 106$, and $M_{wT} = 150$. We cannot match the 90/8/2 product distribution exactly, but we can come close, for example, by operating with x_A between 0.965 and 0.970. Note that this corresponds to conversion of only 3.0–3.5 percent of the water. It is clear that we will be operating with a very large excess of water, which must then be separated from the product.

Table 7.5. *Product distribution for glycols and conversion of water and EtO in a continuous-flow stirred-tank reactor for various conversions of ethylene oxide. (Mass fractions may not sum to unity because of rounding.)*

Water in effluent		Product mass fractions			EtO in effluent
$x_A = \dfrac{c_A}{c_{Af}}$	Conversion of water $1 - x_A$	Monoglycol	Diglycol	Triglycol	$\dfrac{c_{Bf} - c_B}{c_{Af}}$
0.990	0.010	0.966	0.033	0.001	0.010
0.985	0.015	0.950	0.048	0.002	0.016
0.980	0.020	0.934	0.063	0.004	0.021
0.975	0.025	0.918	0.077	0.006	0.026
0.970	*0.030*	*0.902*	*0.090*	*0.008*	*0.032*
0.965	*0.035*	*0.887*	*0.102*	*0.011*	*0.038*
0.960	0.040	0.872	0.115	0.014	0.043
0.955	0.045	0.857	0.126	0.017	0.049
0.950	0.050	0.842	0.137	0.020	0.055
0.945	0.055	0.828	0.148	0.024	0.061
0.900	0.100	0.710	0.221	0.069	0.122
0.850	0.150	0.599	0.267	0.133	0.199
0.800	0.200	0.507	0.289	0.204	0.289
0.750	0.250	0.429	0.294	0.277	0.390

We are now part way through the design process, and we have a considerable amount of information in hand. We presume from this point forward that the desired product distribution has been specified (as before, probably based on a market analysis). The first thing we note is that the amount of water that must be removed per mole of mixed product produced, $c_A/(c_{Af} - c_A) = x_A/(1 - x_A)$, is a constant for a given product distribution. The total production of glycols is given (with a bit of algebra to convert from total molar throughput to mass throughput with a fixed product distribution) by $q(c_{Bf} - c_B)$. We see from Equation 7.19d that the product distribution, which is fixed by x_A, determines only the ratio $(c_{Bf} - c_B)/c_{Af}$, and we are free to choose the actual conversion of B (EtO), c_B/c_{Bf}. Consideration of the results in Table 7.5 give us some immediate insight into the extent to which we should try to convert B and the impact on the resulting process design. From Equation 7.18a we can write

$$\frac{c_B}{c_{Bf}} = \frac{1 - x_A}{x_A} \frac{1}{k_1 c_{Bf} \theta}. \tag{7.20}$$

Now, $1 - x_A$ is a very small number, which we will denote ε. $k_1 c_{Bf}$ has the form of an effective first-order rate constant. If we take $k_1 c_{Bf} \theta$ to be of order unity, then c_B/c_{Bf} will be of order ε. We can thus reach some tentative conclusions about the design: We probably want to choose the residence time to be of order $1/k_1 c_{Bf}$. This will lead to a reactor that is designed for nearly complete conversion of EtO ($c_B/c_{Bf} \sim \varepsilon$), in which case it will not be necessary to have a separate unit to remove unreacted ethylene oxide. (We would most likely set θ to about $3/k_1 c_{Bf}$ or slightly more in order to achieve a conversion of ethylene oxide of 99 percent or better.)

With $c_{Bf} \gg c_B$, it also follows from Table 7.5 that the molar feed ratio of water to ethylene oxide will be about 30 ($c_{Bf}/c_{Af} \sim 0.035$), which is a very large excess relative to the stoichiometric ratio, and a very large amount of water will have to be separated. In this case, the unreacted EtO will be taken off in trace amounts with the unreacted water; the specific numerical coefficient in the selection of the residence time will be based on the maximum concentration of unreacted B that is permitted in the unreacted water stream. The flowrate q will be determined from the required production rate. (The feed can be assumed to be a mixture of pure ethylene oxide and water, so the feed concentration is known.) The reactor volume is then determined from the residence time.

The rigorous solution of the design problem requires that we carry out an analysis similar to the optimal design in the preceding section, including the costs of all separations (products, water, and possibly ethylene oxide), and this can of course be done, but the only relevant question once the product distribution and production rate are set, other than whether the process can show a positive return, is whether we should operate with less than nearly complete conversion of B (ethylene oxide). Complete analyses have been done industrially, and the conclusion is always that the cost of operating with partial conversion and separation of B far outweighs the cost of a larger reactor to achieve nearly complete conversion.

7.5 Concluding Remarks

This is an exceedingly important chapter, in some ways perhaps the single most important chapter in the text in terms of engineering practice. Using the principles developed in the preceding chapters we have arrived at the logical culmination of the analysis process: a practical engineering design. The reactor is clearly the key to the process, since what happens in the reactor affects every other downstream element of the process. The primary point to take away from this chapter is that the equations involving the design variables, such as reactor volume and flow rate, must be combined with the process economics and other constraints in order to obtain a meaningful solution. Furthermore, considerable insight can be obtained with a relatively straightforward analysis based on a few reasonable assumptions.

The design examples in this chapter are quite realistic in form, and the glycol example is, in fact, taken from an actual industrial design study. A "real" design problem will, of course, be more complex. We have assumed here that the reactor will be a single CFSTR; other reactor configurations are possible and must be considered. Reactor temperature and associated energy costs are important variables. We have completely ignored the details of separation. There will often be a larger set of chemical reactions and products, with different selectivity issues. The rate expressions will frequently be more complex than the elementary mass action kinetics applicable to the glycol reactions. The linear cost functions are a gross simplification. *Yet nothing changes in principle*; frequently, only the algebra becomes more difficult, although in many cases considerable computational effort is required. This introduction, if well understood, can provide an intellectual framework for all future studies of the components of a process and the implementation of a process design.

Bibliographical Notes

The continuation and expansion of the subject matter introduced in this chapter may receive some coverage in the core chemical engineering course in kinetics and reaction engineering, and it is touched upon in some textbooks for that course, but in a traditional curriculum the topic is more likely to be covered in a capstone course in process design.

PROBLEMS

7.1. Consider the system studied in Example 7.1. Suppose that the feedstream is specified to be $q = 6 \times 10^{-3}$ m^3/min. Find the production rate qc_M as a function of the reactor volume.

7.2. An irreversible first-order decomposition reaction A $\rightarrow$ M is carried out in a CFSTR, with $k = 0.005$ min^{-1}. The feed composition c_{Af} is 0.2 g-mol/L, and the desired production rate of product is 50 g-mol/min.

  a.   What is the minimum possible flow rate?
  b.   Consider flow rates up to four times the minimum, and calculate the reactor volumes and the effluent concentrations.

7.3. Kinetic data for the reaction between sulfuric acid and diethyl sulfate are given in Examples 6.1 and 6.4. Suppose that $V = 25.4$ L, $q_{Af} = q_{Bf}$, $c_{Af} = 11.0$ g-mol/L, $c_{Bf} = 5.5$ g-mol/L, and the required effluent concentration of sulfuric acid is $c_A = 4.0$ g-mol/L. Find the flow rates (a) assuming that that the reaction may be taken to be irreversible and (b) taking the reverse reaction into account.

7.4. Consider the optimal design problem in Section 7.3, but now suppose that the capital cost of the reactor increases with volume as V^{α}, $\alpha < 1$. Derive the algebraic equation for the optimal conversion, but do not attempt to solve the equation. What is fundamentally different about the solution for $\alpha < 1$?

7.5. The chemical reaction sequence A $\rightarrow$ M $\rightarrow$ S takes place in a CFSTR. You may assume that the reactions are irreversible and first order, with rate constants k_1 for the reaction A $\rightarrow$ M and k_2 for the reaction M $\rightarrow$ S. M is the desired product. Find the residence time $\theta = V/q$ that maximizes the concentration of M in the reactor effluent, and find the maximum concentration.

7.6. The irreversible reaction A $\rightarrow$ products is believed to be nth order in the concentration of A. Devise a strategy for obtaining the order and rate constant by carrying out steady-state experiments in a CFSTR.

7.7. The decomposition of the carcinogen N-nitrosodiethylamine (NDEA) in water was shown in Example 6.3 to be first order when exposed to ultraviolet radiation at an intensity of 1 mW/cm^2, with a rate constant 0.59 min^{-1}.

a. What is the time required in a batch reactor to decrease the concentration in a wastewater sample by a factor of 100?

b. What is the residence time required in a CFSTR to achieve the same reduction?

7.8. Generalize your result to Problem 7.7 as follows: A first-order reaction is to be carried out in a batch reactor and in a CFSTR to achieve a relative reduction RR in concentration. (In Problem 7.7, $RR = 100$.) Compare the required time t_B in a batch reactor to the required residence time θ in a CFSTR to achieve the same RR.

7.9. Follow the approach in Section 5.7 to derive the equation for the dependence of concentration on spatial position in a tubular reactor of length L operating at steady state, in which the radius is R and the volumetric flow rate is q. Assume that there is a single reaction. As in the membrane example, you may assume that there is no mixing in the axial direction, but that the concentration is uniform over any cross-section orthogonal to the direction of flow. How does this equation relate to the equation for the batch reactor?

7.10. The mean residence time in a tubular reactor is the length L divided by the average velocity $v = q/\pi R^2$. Using the results from Problems 7.8 and 7.9, what can you say about the relative values of the mean residence times in tubular and continuous flow stirred-tank reactors to achieve the same conversion for a first-order reaction? Can you generalize this result to an nth order reaction?

8   Bioreactors and Nonlinear Systems

8.1 Introduction

Biotechnology is a major component of modern chemical engineering, and biotechnology appears to many observers to be a new thrust; yet, as noted in Chapter 1, biochemical engineering has been an essential part of chemical engineering since the development of the modern profession in the early part of the twentieth century. One important aspect of biotechnology, in fact its most traditional component, is the use of microorganisms to effect chemical change. We cited the microbiological production of acetone and penicillin as two classic examples in Chapter 1.

    In terms of annual throughput, the activated sludge process for wastewater treatment is by far the most widely used biochemical process in the world, and it provides a useful framework for discussing some interesting features of bioreactor design and performance. The entire process flowsheet is shown schematically in Figure 8.1. The wastewater feed contains organic materials, commonly measured *in toto* as biological oxygen demand (BOD), that are used as nutrients by microorganisms; the organisms produce water and CO_2 as metabolic products. The primary settler is there to remove large objects. The heart of the process is the aeration basin; this is a reactor in which a suspension of microorganisms in porous flocs is brought into contact with the BOD. The microorganisms are aerobic, meaning that they require oxygen for metabolism, so air or enriched air is added to keep the oxygen concentration in the water above a critical level of about 2 g/m^3 (2 ppm). The air jets also serve to mix the reactor contents. The aeration basin and the air feed together account for about 25 percent of the capital cost of the process. The reactor effluent, which contains the suspended microorganisms, goes to a settler, where the microorganisms are separated from the effluent by gravity; the clarified, treated wastewater, with BOD and suspended solids levels below those established by regulatory agencies, is released to the receiving body of water. The microorganisms are then returned to the reactor in a recycle stream.

    What makes the system interesting from a reactor design perspective is that this is an example of an *autocatalytic system*. An autocatalytic system is one in which the

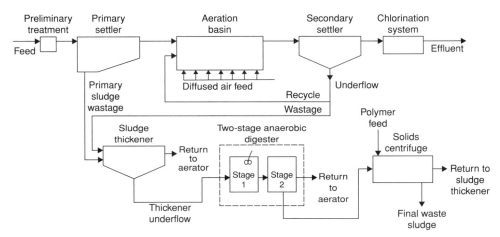

Figure 8.1. Schematic of an activated sludge process.

concentration of one of the reactants is greater at the end of the reaction than at the beginning; that is, the reaction stoichiometry is of the form

$$A + \beta B \rightarrow \gamma B + \text{other products}, \gamma > \beta. \qquad (8.1)$$

In the case of the bioreactor, the amount of biomass grows as the organisms metabolize, so some of the suspended microorganisms (referred to as *sludge*) must be removed before recycling; otherwise, the microorganism content in the aerator would continue to grow until the system clogged. The remainder of the flowsheet is concerned with treatment of the waste sludge, which is thickened by water removal; broken down, usually by anaerobic (no oxygen) bacteria in one or more reactors known as digesters; and dewatered and discarded. (Sludge disposal is itself an environmental problem. Some sludge is dumped in the ocean, and some is used for landfill. Sludge has a high organic content and can be used as fertilizer; there is a commercial market, but heavy metals, which can be concentrated by microorganisms, are a potential hazard.)

8.2 Reactor Analysis

Let us now focus on the reactor-separator system for an autocatalytic reaction. In the case of a bioreactor reactor, such as the activated sludge aerator, we may assume that the slurry containing the suspended microorganisms is a homogeneous medium, and we needn't worry about the fact that two separate phases exist in close proximity. *Homogenization* of this type is a common approach to multiphase systems when the length scale of interest is much larger than the length scale of the microstructure. We can therefore treat both the BOD and the active microorganisms as component species in a continuum.[*] We further assume that the dissolved oxygen level in the reactor is above the critical value of 2 g/m^3, at which level the metabolic rate is insensitive to the O_2 concentration, and that the airflow is sufficient to permit use of the well-mixed assumption. Hence, the reactor can be treated as a single-phase continuous-flow stirred-tank reactor (CFSTR).

In the context of the reaction stoichiometry given in Equation 8.1, we take A to be the BOD and B to be the microorganism. We will presume that all concentrations and rate expressions are in terms of mass units, rather than the molar units commonly employed for reacting systems. The rate of a metabolic reaction is usually of the form

$$r = \frac{\mu c_A}{K + c_A} c_B. \tag{8.2}$$

This form is known as *Michaelis-Menten* kinetics, and is typical of enzymatic reactions. (In the catalysis literature the same functional form is known as *Hougen-Watson* kinetics.) There is a second reaction that the microorganisms undergo, known as endogeneous respiration, which does not depend on the nutrient level, so the rates for BOD and microorganisms differ by more than just the ratio of stoichiometric coefficients, but the effect is small and including endogeneous respiration only adds to the algebraic complexity without changing any essential result. For the activated sludge system, μ is typically about 5 day^{-1} and K is about 100 g/m^3 of BOD$_5$. (BOD$_5$ is a particular measure of the organic level.) The typical feedstream will contain about 200 g/m^3 of BOD$_5$. If we assume that the conversion is at least 95 percent, we may assume that the concentration in the reactor is of order 10 g/m^3, so $c_A \ll K$. In that case we can neglect the c_A term in the denominator and write the rate in the form

$$r = kc_A c_B, \tag{8.3}$$

where $k = \mu/K$. (This approximation is not necessary for any of the analysis that follows, but it is reasonable and does simplify the algebra. Everything can be repeated, with equivalent results, for the full Michaelis-Menten kinetics.)

The system of interest is shown schematically in Figure 8.2. The reactor is taken as the control volume. The steady-state equations for the reactor, with the usual assumptions about the density, and with the stoichiometry given by Equation 8.1 and the reaction rate given by Equation 8.3, are

$$\text{A: } 0 = q(c_{Af} - c_A) - Vkc_A c_B, \tag{8.4a}$$

$$\text{B: } 0 = q(c_{Bf} - c_B) + (\gamma - \beta)Vkc_A c_B. \tag{8.5}$$

There is no B in the external feed to the reactor, but a fraction f of the mass of species B in the reactor effluent is recycled to the feedstream; hence,

$$qc_{Bf} = fqc_B. \tag{8.6}$$

---

[*] Transport of the oxygen through the floc does affect the apparent rate of reaction. Similar issues are discussed in Section 10.5. For our purposes here, however, we can assume that this effect is already incorporated in the reaction rate.

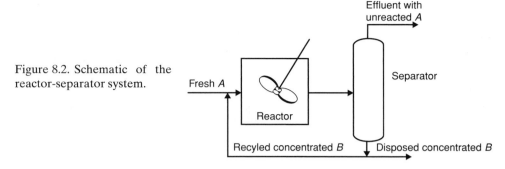

Figure 8.2. Schematic of the reactor-separator system.

Equation 8.5 therefore becomes

$$\text{B}: 0 = qc_B(f-1) + (\gamma - \beta)Vkc_Ac_B = c_B[q(f-1) + (\gamma - \beta)Vkc_A]. \quad (8.4b)$$

Note that $c_B = 0$, $c_A = c_{Af}$ is always a solution to Equations 8.4a and b. Since we are interested in solutions for which $c_B \neq 0$, we must define the range of such solutions. If we do find solutions with $c_B > 0$ (and we will), we must deal with the fact that the process equations admit two physically permissible solutions ($c_B > 0$, $c_B = 0$), and we will have to determine which state can actually be attained.[*]

With $c_B \neq 0$, Equation 8.4b can be solved for c_A:

$$c_B \neq 0 : c_A = \frac{q(1-f)}{(\gamma - \beta)Vk}. \quad (8.7a)$$

Note that f must be strictly less than unity; for $f = 1$, $c_A = 0$, in which case $c_B \to \infty$ in order to satisfy Equation 8.4b. This is consistent with our physical understanding: If all of the B that is produced is returned to the reactor, then the amount of B will continue to grow without bound as long as there is sufficient A in the reactor for the reaction to occur.

We now solve Equation 8.4a to obtain

$$c_B = \frac{(\gamma - \beta)c_{Af}}{1-f} - \frac{q}{Vk}. \quad (8.7b)$$

This is a difference between two positive numbers. Since we require $c_B > 0$, Equation 8.7b places a limit on the reactor variables, as follows:

$$c_B > 0 \Rightarrow \frac{q}{V} < \frac{(\gamma - \beta)kc_{Af}}{1-f}. \quad (8.8)$$

q/V is known in the environmental and biochemical engineering literature as the *dilution rate*; it is the reciprocal of the residence time, θ. If the dilution rate exceeds the bound set by Equation 8.8, then the only physically meaningful solution that can exist to Equations 8.4a and b is $c_B = 0$, $c_A = c_{Af}$. Such as situation is known as *washout*. The restriction given by Equation 8.8 demonstrates the need for good flow

---

[*] The notion that nonlinear equations can have more than one solution is not surprising. Consider the equation $x^2 = 1$, which has solutions $x = 1$ and $x = -1$. It has probably been your past experience that one solution can be eliminated on physical grounds; an absolute temperature cannot be negative, for example. That is not the case here.

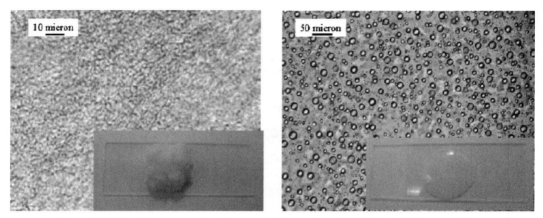

Figure 8.3. Bistable states of a 10 percent dispersion of 8CB in polydimethylsiloxane. From Y. W. Inn and M. M. Denn, *J. Rheology*, **49**, 887–895 (2005).

control in autocatalytic systems such as the activated sludge process, since excursions in flow rate or volume leading to long-term violations of Equation 8.8 could take the system irreversibly to the washout state.

8.3 Nonlinearity

Nonlinear systems often exhibit multiple solutions, and it is quite common for more than one solution to fall within a physically acceptable range. Indeed, more than one steady state may be attainable in practice, depending on the starting conditions. This situation is sometimes called *bistability*, indicating that each state is stable within its immediate environs, but the system might jump to the other state if perturbed sufficiently. Bistability is a common theme in materials processing, where great care must often be taken to ensure that the proper microstructure is achieved. It is also common in chemical and biochemical reaction engineering, microfluidics, and many other applications area. The example in the preceding section illustrates a rather elementary example, although one of extreme practical importance.

Figure 8.3 shows an interesting example of bistability in materials processing, in which a liquid crystal is dispersed in a polymer matrix. Liquid crystals are melts or solutions made up of rigid molecules with large aspect ratios. In some temperature or concentration ranges the molecules align, despite the fact that they are in a liquidlike state, which permits polarized light to pass only with certain orientations. The orientations respond to local external electromagnetic fields, so images can be formed. This, in brief, is the basis of liquid crystal displays. The figure shows a 10 percent by weight dispersion of 4′-octyl-4-biphenylcarbonitrile (commonly known as 8CB) in polydimethylsiloxane. The large images are optical micrographs of thin samples, whereas the inserts are images of the bulk dispersions. Note the different scales on the micrographs. The sample on the left is a fairly rigid gel, with no obvious length scale for the dispersed phase at this level of magnification, whereas the sample on the right is a mobile liquid with a dispersed-droplet morphology. The

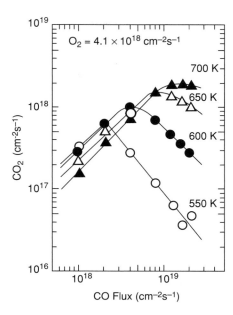

Figure 8.4. Rate of production of CO_2 over a palladium catalyst as a function of available CO. Reproduced from H. Uetsuka, K. Watanabe, H. Kimpara, and K. Kunimori, *Langmuir*, **15**, 5795–5799 (1999) with permission of the American Chemical Society.

only difference between the two is the temperature at which the sample was initially prepared, and both states exist in the temperature range 33–41°C. The gel can be converted to the dispersed droplet morphology by raising the temperature above 41°C and then reducing the temperature, while the dispersed droplet morphology can be converted to the gel by lowering the temperature below 33°C and applying a lot of shear. Shear at a higher temperature will not cause the transition.

Multiplicities resulting from nonlinearities in reacting systems usually require consideration of the effect of temperature on reaction rates and incorporation of the energy balance, which requires concepts that we have not yet addressed. There is one rather nice example that illustrates multiplicity in a reacting system in which the states all have a finite conversion that can be addressed using only tools that we have developed to this point, and we consider that example in the following section.

8.4 CO Oxidation

Figure 8.4 shows data for the rate of oxidation of carbon monoxide as a function of available CO in the presence of a palladium catalyst. The data are expressed in terms of fluxes because of the nature of the reactor that was employed. They clearly show very unexpected behavior, namely that the rate of oxidation initially increases with increased availability of CO, but then the rate goes through a maximum and decreases. This type of rate expression is not common, but it is observed in other systems as well. The mechanism is well understood, but it is not important to us here and we will not pursue it.

Now, suppose that we could carry this reaction out in a CFSTR. To do so in practice would require that we have the catalyst uniformly suspended in the reacting stream such that we could consider the gas-solid suspension as a continuum, as we

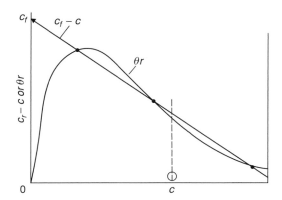

Figure 8.5. Left- and right-hand sides of Equation 8.10 plotted versus CO concentration, showing three solutions with positive concentrations.

did in the preceding example with the liquid-sludge suspension. We will make that assumption. (A fluidized bed would approximate this situation, although a fluidized bed would probably not be the preferred reactor configuration for this conversion.) We will also assume that CO is present in very small quantities in the feedstream, so that we can ignore the density change that would result from the different total number of moles in the feed and effluent at constant temperature and pressure.

The equation for the concentration of CO in a CFSTR is

$$\theta \frac{dc}{dt} = c_f - c - \theta r, \tag{8.9}$$

where we assume that oxygen is available in excess and the only concentration that is relevant for the reaction is the concentration of CO, denoted simply as c. At steady state we can write

$$c_f - c = \theta r. \tag{8.10}$$

The right side of Equation 8.10, the dimensionless reaction rate, is given as a function of c by a curve with the general form of the data in Figure 8.4, and is plotted schematically in Figure 8.5. The left side of Equation 8.10 is a straight line with slope -1, and it is also plotted in Figure 8.5. The two curves can intersect three times, meaning that *there are three possible concentrations with $0 < c < c_f$ for which Equation 8.4 is satisfied for a given residence time and feed concentration*, corresponding to low, intermediate, and high conversions. The concentrations are all physically realistic.

It is possible to show, by detailed analysis of the nonlinear differential Equation 8.9 (see Appendix 8A for a partial treatment), that the intermediate solution cannot be maintained in practice; that is, it is a solution to the steady-state model equation, but it is *unstable* in that any infinitesimal disturbance (which cannot be avoided in practice) will cause the system to move to one of the other steady states. The high- and low-conversion steady states are both stable, however, and the system can operate at either in the absence of large disturbances that may cause it to move to the other state. This type of behavior is characteristic of highly energetic systems that can ignite; it is also a characteristic of some model biological systems that can switch between states. The major message to carry away is that multiplicity is to be expected, and the possibility should always be examined in any physical process.

8.5 Concluding Remarks

Physicochemical systems, which are inherently nonlinear, frequently exhibit multiple solutions corresponding to multiple states. The examples illustrated analytically here, namely finite conversion and washout in a bioreactor and low and high conversions in CO oxidation, are good case studies because they represent important processes for which the behavior can be elucidated with relatively straightforward descriptions. Reactor multiplicities usually involve thermal effects, and one example is addressed in Chapter 15; the subject is typically covered in the chemical engineering curriculum in a course with a title along the lines of Kinetics and Reactor Design. Multiplicities involving phase changes – the liquid crystal gel/dispersed droplet bistability, for example, or a supersaturated single phase when precipitation and the formation of a second phase would be expected – normally require a detailed thermodynamics analysis using concepts addressed in a course with a title such as Thermodynamics or Phase Equilibrium. The main point to carry away from this chapter, beyond the fact that bioreactors are simply autocatalytic systems that require special care because the catalyst is a living entity, is that one must always be aware of the possibility of instabilities and multiplicities in nonlinear systems, because the consequences of going to a state other than the desired one can be serious and sometimes catastrophic.

Bibliographical Notes

The activated sludge process is the prototypical wastewater process, and it provides an excellent introduction to the notion of washout and the more general subject of multiplicity in physicochemical systems. The broad subject of wastewater treatment is covered in

> Metcalf and Eddy, Inc., *Wastewater Engineering: Treatment and Reuse*, 4th Ed., McGraw-Hill, New York, 2003.

Two recent reviews that address applications and especially reactor and process configurations, but not the issues that are the focus of this chapter, are

> Chan, Y. J., M. F. Chong, C. L. Law, and D. G. Hassell, "A review on anaerobic-aerobic treatment of industrial and municipal wastewater," *Chemical Engineering Journal*, **155**, 1–18 (2009).
> Kassab, G., M. Halalsheh, A. Klapwijk, M. Fayyad, and J. B. van Lier, "Sequential anaerobic-aerobic treatment for domestic wastewater – A review," *Bioresource Technology*, **101**, 3299–3310 (2010).

Autocatalysis is an important element in the self-replication of biological systems, and much of the recent literature addresses its role in the development of the cell. One review that is readable and gives some insight into the scientific issues is

Loakes, D., and P. Hollinger, "Darwinian chemistry: towards the synthesis of a simple cell," *Molecular Biosystems*, **5**, 686–694 (2009).

Some autocatalytic reactions can exhibit sustained oscillations in a region of the parameter space. One of the earliest theoretical studies, which ultimately led to the *Lotka-Volterra model* for predator-prey interactions in population dynamics, is accessible through the American Chemical Society online journal archives:

Lotka, A. J. "Contribution to the theory of periodic reactions," *J. Physical Chemistry*, **14**, 271–274 (1910).

The article is quite readable and highlights the interesting material that can often be found in the original literature on a subject.

Nonlinearity permeates modern science and engineering, and the examples in this chapter simply give a taste. Further examples will arise in a natural way in the core course in thermodynamics, but broader issues generally require mathematics that goes beyond what is typically covered in a first course in ordinary differential equations. A brief introduction to multiplicity in nonisothermal chemical reactors, which is related to ignition and extinction in combustion, will be found in Chapter 15. The popular press often refers to the subject of *chaos*, which is a manifestation of nonlinearity in systems that evolve in time. A discussion at a popular level can be found in

Gleick, J., *Chaos: Making a New Science*, Penguin, New York, 1988, 2008.

Appendix 8A: Dynamical Response for CO Oxidation

We can obtain some insight into the dynamical response of the CO oxidation reactor, and see that the intermediate steady state cannot be maintained in practice, by some elementary mathematics. The reaction rate to the right of the maximum in Figure 8.4 behaves roughly as c^{-1}. If we then set

$$r = kc^{-1}, \tag{8A.1}$$

recognizing that this form is not even qualitatively correct at low concentrations, the dynamical response, Equation 8.9, is given by

$$\theta \frac{dc}{dt} = c_f - c - \frac{k\theta}{c}. \tag{8A.2}$$

This equation has two steady states, which correspond to the middle and upper steady states in Figure 8.4. The steady states, denoted c_s, are then given by

$$c_s = \frac{c_f \pm \sqrt{c_f^2 - 4k\theta}}{2}. \tag{8A.3}$$

The middle and upper steady states will no longer exist if $\theta > c_f^2/4k$ (c_s becomes complex).

It is convenient (and very useful) in analyzing dynamics to use the steady state as a frame of reference; that is, we define a new dependent variable

$$\xi = c - c_s. \tag{8A.4}$$

We will consider each of the steady states in turn, so c_s denotes either of the two solutions in Equation 8A.3, as appropriate. Note that $d\xi/dt = dc/dt$, since $dc_s/dt = 0$. Equation 8A.2 then becomes

$$\theta \frac{d\xi}{dt} = c_f - c_s - \xi - \frac{k\theta}{c_s + \xi} = \left(c_f - c_s - \frac{k\theta}{c_s} \right) - \xi$$
$$- k\theta \left(\frac{1}{c_s + \xi} - \frac{1}{c_s} \right) = -\xi \left(1 - \frac{k\theta}{c_s^2 + c_s\xi} \right). \tag{8A.5}$$

The first grouping in parentheses after the second equality in Equation 8A.5 sums to zero because of the steady state equation (it is simply the right-hand side of Equation 8A.2 when $dc/dt = 0$.)

Equation 8A.5 is separable (i.e., we can write the equation in the form $\int d\xi/f(\xi) = -\int dt/\theta$), and the integration can be done analytically, but the result is a transcendental function that is not easy to interpret, so we will instead make the approximation $|\xi| \ll c_s$ (i.e., the concentration is always very close to the steady state concentration). In that case Equation 8A.5 can be written approximately as

$$\theta \frac{d\xi}{dt} = -\xi \left(1 - \frac{k\theta}{c_s^2} \right), \tag{8A.6}$$

which is a separable equation with the solution

$$\xi = \xi_0 \exp \left[- \left(1 - \frac{k\theta}{c_s^2} \right) \frac{t}{\theta} \right]. \tag{8A.7}$$

Thus, for $k\theta/c_s^2 < 1$, which corresponds to the high concentration/low conversion steady state (the positive sign in Equation 8A.3), $\xi(t)$ always goes to zero after long times, which means that c always returns to c_s after a disturbance (as long as the magnitude of the disturbance ξ_0 is small enough to permit the approximation that was made). This steady state can therefore be maintained in practice, and it is *stable* to sufficiently small disturbances. By contrast, for $k\theta/c_s^2 > 1$, which is the middle steady state in Figure 8.4 (the negative sign in Equation 8A.3), Equation 8A.7 is a growing exponential and $\xi(t)$ grows without bound for arbitrarily small ξ_0. Thus, the system will *always* move away from this steady state, even with starting concentrations that are arbitrarily close, and it is *absolutely unstable*; that is, it can never be maintained in practice. (The fact that $\xi(t)$ always moves away from ξ_0 and from zero is the important point here, not that it grows without bound; clearly, at some point in this growth process, the assumption that $|\xi| \ll c_s$ will fail, and Equation 8A.6 will no longer describe the physical process.)

9 Overcoming Equilibrium

9.1 Introduction

Most chemical reactions are limited by equilibrium, and in many cases the equilibrium constant is such that the conversion to the desired product is too small to be economical. Doing "better than equilibrium" is one of the primary challenges in chemical engineering design. The general idea is obvious: The reaction should be carried out in the absence of product, so that the reverse reaction cannot proceed. Implementation is not so obvious, which is why this is a primary challenge.

The most interesting approach to overcoming equilibrium, which has been successful in many cases, is to design a system in which reaction and product separation take place simultaneously. In that way, the product is removed from the reactor and cannot participate in the reverse reaction. The inspiration for this idea may come from the biological cell, in which the reaction sites are enclosed within the cell membrane, which is permeable to reaction products that are intended for use outside the cell. The two most common manifestations outside nature are the membrane reactor, which mimics the cellular process, and reactive distillation, in which the reaction takes place in a distillation column. A proper treatment of distillation requires inclusion of the energy balance and the notion of vapor-liquid equilibrium, which is typically addressed in a subsequent course in thermodynamics, so we will not examine reactive distillation here. The membrane reactor is accessible to us now, however, and nicely illustrates the concept.

9.2 Equilibrium-Limited Continuous-Flow Stirred-Tank Reactor

To establish a frame of reference, let us consider the reversible reaction $A \rightleftarrows M$ in a steady-state CFSTR. We will assume that the rates of both the forward and reverse reactions are first order. (As in many other examples throughout this text, nothing important changes qualitatively if we consider reactions with more realistic kinetics, but the important ideas become lost in the algebra.) We will assume that there is no product in the feed to the reactor. The equations for the two component species are then

$$A: 0 = c_{Af} - c_A - k_1\theta c_A + k_2\theta c_M, \tag{9.1a}$$

$$M: 0 = -c_M + k_1\theta c_A - k_2\theta c_M. \tag{9.1b}$$

By adding the two equations we obtain $c_M = c_{Af} - c_A$, and we then obtain the effluent concentration of A from Equation 9.1a as

$$\frac{c_A}{c_{Af}} = \frac{1 + k_2\theta}{1 + (k_1 + k_2)\theta}. \tag{9.2}$$

The conversion is less (i.e., c_A is larger) when $k_2 > 0$ (reversible) than when $k_2 = 0$ (irreversible). [The result is readily established by showing that $\partial(c_A/c_{Af})/\partial k_2 > 0$, which is a straightforward calculation that is worth doing.] The limiting case of $\theta \to \infty$, in which the reactions have time to proceed to equilibrium, gives

$$\theta \to \infty: \frac{c_A}{c_{Af}} = \frac{1}{1 + K_{eq}}, \tag{9.3}$$

where $K_{eq} = k_1/k_2$ is the equilibrium constant for the reaction (the ratio c_M/c_A at equilibrium).

9.3 Single-Stage Membrane Reactor

Now consider the scheme shown in Figure 9.1. We have a single-stage membrane separation system, identical to the one in Figure 5.1, but now we permit the reversible reaction $A \rightleftarrows M$ to occur on the raffinate side of the membrane. Only the product, M, can cross the semipermeable membrane. We retain the symbols R and P for the flow rates of the raffinate and permeate stream, respectively. A is present only in the raffinate stream, so we can retain the symbol c_A for the concentration of A, but we need to include subscripts R and P to distinguish the concentrations c_{RM} and c_{PM} of M in the raffinate and permeate streams, respectively. We assume that there is no M present in the permeate feed.

As in Section 5.2, we now require two distinct control volumes, separated by the membrane. We must consider two rates when writing the component mass balance equation for M on the raffinate side of the membrane: the net rate of reaction

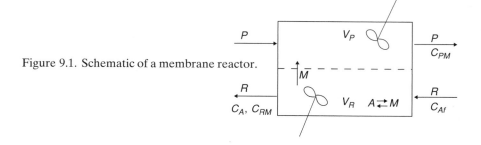

Figure 9.1. Schematic of a membrane reactor.

(forward and reverse) and the rate of transfer of M across the membrane. The steady-state equations for A in the raffinate stream and M in the raffinate and permeate streams, respectively, are

$$\text{Raffinate, A:} \quad 0 = R(c_{Af} - c_A) - k_1 V_R c_A + k_2 V_R c_{RM}, \tag{9.4a}$$

$$\text{Raffinate, M:} \quad 0 = -R c_{RM} + k_1 V_R c_A - k_2 V_R c_{RM} - \Pi A(c_{RM} - c_{PM}), \tag{9.4b}$$

$$\text{Permeate, M:} \quad 0 = -P c_{PM} + \Pi A(c_{RM} - c_{PM}). \tag{9.4c}$$

As in Chapter 5, Π denotes the membrane permeability and A denotes the available membrane area. (We have a conflict in nomenclature here, where we are using A with two slightly different typefaces to denote both the reactive species and the membrane area, respectively. There should be no confusion, since the symbol that appears in the equations refers only to the area.)

The algebra to solve this system of three linear equations is tedious, but straightforward. We first find c_{PM} in terms of c_{RM} from Equation 9.4c,

$$c_{PM} = \frac{\Pi A}{P + \Pi A} c_{RM}. \tag{9.5}$$

We then use this result in Equation 9.4b to obtain c_{RM} in terms of c_A:

$$c_{RM} = \frac{k_1 V_R}{R + k_2 V_R + P \dfrac{\Pi A}{1 + \Pi A}} c_A. \tag{9.6}$$

Finally, we obtain c_A from Equation 9.4a:

$$\frac{c_A}{c_{Af}} = \frac{1 + k_2 \theta}{(1 + k_1 \theta)\left(1 + \dfrac{P}{R}\dfrac{\Pi A}{1 + \Pi A}\right) + k_2 \theta}. \tag{9.7}$$

Here, $\theta = V_R / R$.

Equation 9.7 reduces to Equation 9.2 if there is no membrane transport (P, Π, or A goes to zero). It reduces to the result for a single-stage membrane unit in Section 5.2 if there is no reaction (k_1 and k_2 both go to zero), although a bit of algebra is required to obtain the equivalent of Equation 5.4 for c_{RM}. It is obvious by inspection that the conversion of A given by Equation 9.7 is greater than that given by Equation 9.2 for equal values of the residence time in the reactor. The potentially dramatic effect of removing the reaction product can best be seen by considering the case in which the residence time is large ($k_1 \theta \gg 1$, $k_2 \theta \gg 1$), in which case Equation 9.7 simplifies to

$$\theta \to \infty: \quad \frac{c_A}{c_{Af}} = \frac{1}{1 + \left(1 + \dfrac{P}{R}\dfrac{\Pi A}{1 + \Pi A}\right) K_{eq}}. \tag{9.8}$$

Clearly, the larger the membrane transport term, the greater the improvement will be on equilibrium.

9.4 Concluding Remarks

There is no need that a membrane reactor be a single-stage device, nor is it apparent that it should consist of only one stage. We have seen in Chapter 5 how cross-flow and countercurrent flow configurations for membrane separation systems can enhance the separation efficiency. The same configurations can be used for membrane reactors, and in fact the typical configuration for a membrane reactor is as a counterflow device, usually with a ceramic membrane. Membrane reactors are also commonly used for biochemical conversions in which cells are placed on one side of a tubular membrane and product flows across the membrane. We shall not pursue the calculations for multistage reactors here; the calculations are straightforward but involve a great deal of algebra.

The main point to take away from this short chapter is that clever exploitation of rate processes can greatly enhance process performance, and that asking the correct question (in this case, "What would increase the conversion?") is the most important step in getting a useful answer.

Bibliographical Notes

For practical discussions of membrane reactors, see

Tsotsis, T. T., A. M. Champagnie, S. P. Vasileiadis, Z. D. Ziaka, and R. G. Minet, "The enhancement of reaction yield through the use of high-temperature membrane reactors," *Separation Science and Technology*, **28**, 397–422 (1993).
Buxbaum, R. E., "Membrane reactor advantages for methanol reforming and similar reactions," *Separation Science and Technology*, **34**, 2113–2123 (1999).

A discussion of reactive distillation, with references to the first studies, can be found in

Dudukovic, M. P., "Challenges and innovations in reaction engineering," *Chemical Engineering Communications*, **196**, 252–266 (2009).

PROBLEMS

9.1. In Equation 9.2, show that $\partial(c_A/c_{Af})/\partial k_2 > 0$.

9.2. In analogy to the treatment in Section 5.7, derive the equations describing a countercurrent membrane reactor in which the reversible first-order reaction $A \rightleftarrows M$ is carried out.

10 Two-Phase Systems and Interfacial Mass Transfer

10.1 Introduction

Many physicochemical systems consist of two or more phases in intimate contact. We have already considered one such system in Chapter 8, where we noted that the bioreactor contains flocs of microorganisms suspended in an aqueous phase, together with air bubbles that provide oxygen. In that case, we made the approximation that we could treat the system as though it were one continuous phase, a process known as *homogenization*. We did the same for the CO oxidation reactor with finely dispersed catalyst. Homogenization works when all microstructural length scales are so small that the phenomena occurring in the various phases can be averaged together over a continuum length scale that is still very small relative to the macroscopic size of the system, in which case any time scales associated with transport within the microstructure are negligible relative to overall system scales. In many multiphase cases, however, the length scales are such that we must directly address the multiphase nature of the system and the concomitant transport of mass and energy between the phases.

The focus of this chapter will be on the foundations of interfacial mass transfer of a component species between the phases and the approach to an equilibrium distribution. The following chapter will address some of the processing issues that arise. As a preface to the analysis, however, it is useful to begin with a brief classification of various types of two-phase systems that may be encountered in practice.

10.2 Classification of Two-Phase Systems

Substances typically occur in any of three states – solid, liquid, or gas – in applications of interest to chemical engineers. (Plasmas and liquid crystals are often considered to be states distinct from the classical trio, but we shall not pursue these here.) For engineering purposes, we define a *phase* as a macroscopic portion of a system that is composed entirely of material in one state and that has an identifiable interface with the other phase or phases of the system. Thus, we encounter two-phase systems that are solid-liquid, solid-gas, and liquid-gas. Immiscible liquids also form distinct phases, and liquid-liquid systems are common in processing applications. Solid-solid systems are not often encountered in the applications with which we will be concerned, and gas-gas systems cannot exist, since mixtures of gases do not form distinct phases.

A two-phase system may occur because of a processing operation in which the raw materials exist in distinct phases, it may result from the deliberate contacting of one phase with another, or it may be the consequence of a second phase being produced during a process. For classification purposes, it is convenient to distinguish between processes where a significant phase change occurs and those where it does not. We shall deal only with the latter situation in this text. In this chapter and the next we consider processes where two phases are contacted for the purpose of interfacial mass transfer.

Most mass transfer processes are similar in concept, implementation, and quantitative description. As technology has developed, a variety of names have been assigned to these processes, commonly referred to as *unit operations*. These operations are briefly described here.

Solid-liquid systems. Solid-liquid mass transfer systems are of two types. In *adsorption,* dissolved or suspended material is transferred from the liquid phase to the surface of the solid. Activated charcoal, silica gel, and magnesium oxide are typical adsorbents for applications such as the removal of sulfur compounds from gasoline, of water from hydrocarbons, and of various impurities from water. *Ion exchange* is like adsorption except that an ion transferred to the solid surface is replaced in the fluid by an ion of the same charge that is removed from the solid. Ion exchange is used, for example, in water treatment and for purification of pharmaceuticals. Polymeric resins such as crosslinked polystyrene and inorganic zeolites (hydrated alumino-silicates) are typical ion exchange solids.

Solvent extraction is the transfer of a soluble material from a solid phase to a liquid solvent. The term *washing* is often used when the solvent is water and the solid to be removed is adhering to the surface of an insoluble solid. More complex separations are called *leaching.* Typical applications are the recovery of copper from low-grade copper oxide ores by extraction with sulfuric acid and the leaching of sugar from beet pulp with water.

Solid-gas systems. The transfer of material from a gas to a solid surface is also called adsorption. A typical example is the use of silica gel for the removal of SO_2 and water from gas mixtures. Adsorption on a solid surface is also an essential step in the use of solid catalysts for chemical reactions of gaseous species. *Drying,* or *desorption,* is the transfer of a volatile substance bound in the solid phase to a gas stream.

Liquid-gas systems. Transfer of a component from a gas phase to a liquid is known as *absorption,* or sometimes as *scrubbing.* (*Absorption* and *adsorption* sound very much alike, and one is often used incorrectly in place of the other.) Removal of SO_2 from air by absorption into aqueous sodium carbonate and removal of H_2S from natural gas with monoethanolamine solution are typical applications with significant environmental importance. Absorption is also a key step when a gaseous reactant is brought into contact with a liquid phase reactant, as in the bioreactor discussed in Chapter 8. In that case, the oxygen necessary for the bio-oxidation reactions that break down the organic matter is supplied by bubbling air through the liquid phase. *Desorption,* or *stripping,* are terms used when a component of the liquid is transferred to the gas, as in the removal of CS_2 from an oil phase using steam.

Liquid-liquid systems. Solvent extraction is the transfer of a solute from one liquid phase to another when the two solvents are themselves immiscible (or, at least, nearly so). Because the two phases are both liquids, the separation following intimate contact of the two phases for purposes of solute transfer must be followed by a step in which the two liquids are separated; this is usually done through settling enabled by density differences. Typical uses of extraction are in the recovery of penicillin from the fermentation broth using cyclohexane or chloroform as a solvent and the dewaxing of lubricating oils using ketones or liquid propane.

The unit operations described here all involve intimate contact between two phases to enable the transfer of material from one phase to the other. All can be carried out isothermally. The different types of materials involved in the various unit operations necessitate that the process equipment for effecting the mass transfer must differ in each case, but the basic principles are essentially the same and it will be possible to consider the basics of the mathematical description of the mass transfer simultaneously for all.

Operations involving a phase change require further consideration because the principle of conservation of energy is generally required, and we leave these for subsequent study. The most common separation process requiring phase change is *distillation,* in which a liquid mixture is boiled to produce a vapor of different composition. Repeated systematic application leads to purification. In *crystallization,* a solution is cooled to produce a solid phase of different composition. Despite the different physical bases, aspects of these phase-change operations are described with mathematical models quite similar to those needed for the mass transfer unit operations described above, and the processing concepts developed in the next chapter will apply to them as well.

10.3 Batch Two-Phase Systems

10.3.1 Basic Model Equations

A batch process is one in which there is no flow into or out of the system. As we saw in Chapter 6, batch processes can be useful for determining reaction rate information. They are also useful for determining rates of mass transfer. As with the batch reactor, we charge the two phases to the stirred vessel at time $t = 0$. We assume that the two phases are in intimate contact, with one phase uniformly distributed throughout the other; uniformity is easy to achieve in practice for many systems. We further assume that each phase is individually well mixed, so that a sample of either phase drawn at any time will be the same as any other sample of that phase drawn

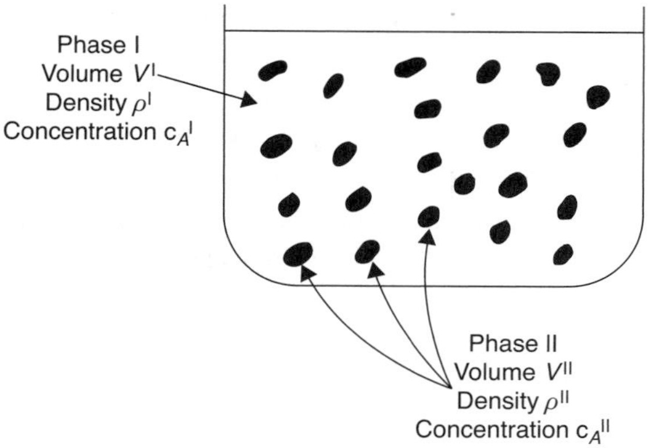

Phase I
Volume V^I
Density ρ^I
Concentration $c_A{}^I$

Figure 10.1. Schematic of a two-phase system. Each phase must be taken as a separate control volume.

Phase II
Volume V^{II}
Density ρ^{II}
Concentration $c_A{}^{II}$

from any position in the tank at the same time. We have already seen the importance of the assumption of perfect mixing for single-phase systems, and the experimental achievement of near-perfect mixing for each phase is of comparable importance in our analysis here. We assume that the mass transfer occurs isothermally; this is frequently a valid assumption, and the isothermal analysis is always part of the more complete treatment when thermal effects do have to be taken into account.

As before, we will use volume V, density ρ, and concentration c as characterizing variables for mass. It is conventional when considering interfacial mass transport to measure concentration in mass units (e.g., kg/m^3 or lb$_m$/ft^3), in contrast to the use of molar units when we were considering reacting systems. This practice causes no inherent difficulty, but it does require some care when analyzing systems in which there is both interfacial mass transfer and chemical reaction, since the reaction rates will be expressed most naturally in molar units. Selection of the control volume requires some care. It is clear that the entire vessel is not a useful control volume for the two-phase system in most cases. We are interested in transfer *between* the two phases, hence we must work with *two* control volumes, one for each phase. We will arbitrarily designate one phase as the *continuous* phase (Phase I) and one as the *dispersed* phase (Phase II). The continuous phase consists of the Swiss cheese-like volume shown in Figure 10.1; the volume of the continuous phase is denoted V^I, the density ρ^I, and the concentration of any species i is denoted c_i^I. The volume of the dispersed phase, V^{II}, is made up of all the elements of the other phase and, although we treat it as a single volume for modeling purposes, it may consist physically of a number of distinct volumes. The density of the dispersed phase will be denoted by ρ^{II}, and the concentration of any species i is denoted c_i^{II}.

We will develop the mathematical description for a general two-phase system, recognizing that with appropriate identification of the continuous and dispersed phases, the basic model equations will apply for batch solid-liquid, solid-gas, liquid-liquid, or liquid-gas systems as long as the assumption of density and concentration uniformity within each phase can be maintained. (Meeting this requirement may be difficult in systems with a solid phase, or systems with a very viscous liquid, where transport *within* the phase may be very slow.)

To simplify the algebraic manipulations and emphasize the physical processes, we will develop the model equations for the case in which a single component species A is transferred between the phases. A has a concentration c_A^I in the continuous phase and c_A^{II} in the dispersed phase and does not react with any component in either phase. It is not difficult conceptually to extend the treatment to any number of species and to include reaction in one or both phases; in fact, we will introduce chemical reactions later in this chapter.

The equations of conservation of mass will require an expression for the rate at which each component species is transferred between the phases. We will use the boldface symbol $\mathbf{r}$ to distinguish this rate from the rate of chemical reaction. Thus, the rate at which A enters Phase I through mass transfer across the interface is denoted $\mathbf{r}_{A+}^I$, whereas the rate at which A is depleted in Phase I by mass transfer across the interface is denoted $\mathbf{r}_{A-}^I$. Similarly, the rate of accumulation of A in Phase II through interfacial mass transfer is $\mathbf{r}_{A+}^{II}$, and the rate at which A is lost from Phase II through interfacial mass transfer is $\mathbf{r}_{A-}^{II}$. The dimensions of the rate of mass transfer are mass per area per time. The rate is written on a per-area basis because, all other things being held equal, an increase in the area between the phases will lead to a proportionate increase in the mass transferred. (We have already utilized this concept in Chapter 5 in writing the rate of solute transfer through the membrane on an area basis.)

Let a denote the total interfacial area between the phases. The equations for conservation of mass in each of the control volumes are then, respectively,

$$\frac{d\rho^I V^I}{dt} = a[\mathbf{r}_{A+}^I - \mathbf{r}_{A-}^I], \tag{10.1I}$$

$$\frac{d\rho^{II} V^{II}}{dt} = a[\mathbf{r}_{A+}^{II} - \mathbf{r}_{A-}^{II}]. \tag{10.1II}$$

It is evident for the batch system that the mass that leaves Phase I must go to Phase II, and vice versa. Thus, $\mathbf{r}_{A+}^I = \mathbf{r}_{A-}^{II}$ and $\mathbf{r}_{A-}^I = \mathbf{r}_{A+}^{II}$. For convenience, we define the net rate r_A as

$$\mathbf{r}_A = \mathbf{r}_{A+}^{II} - \mathbf{r}_{A-}^{II}. \tag{10.2}$$

Then

$$\frac{d\rho^I V^I}{dt} = -a\mathbf{r}_A, \tag{10.3I}$$

$$\frac{d\rho^{II} V^{II}}{dt} = a\mathbf{r}_A. \tag{10.3II}$$

Application of conservation of mass to the species that is transferred, component A, leads in an identical manner to the component equations:

$$\frac{dc_A^I V^I}{dt} = -a\mathbf{r}_A, \tag{10.4I}$$

$$\frac{dc_A^{II} V^{II}}{dt} = a\mathbf{r}_A. \tag{10.4II}$$

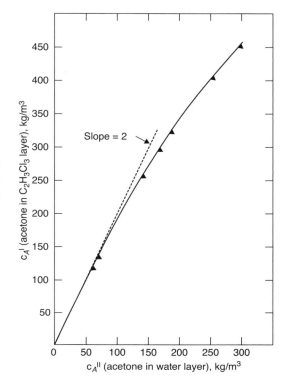

Figure 10.2. Equilibrium concentration of acetone in 1,1,2-trichloroethane at 25°C as a function of equilibrium concentration of acetone in water. Data of R. E. Treybal, L. E. Weber, and J. F. Daley, *Ind. Eng. Chem.*, **38**, 817–821 (1946).

10.3.2 Rate Expression

As with a reacting system, it is necessary to establish the constitutive equation relating the rate $\mathbf{r}_A$ to the other system variables, particularly the concentrations c_A^I and c_A^{II}. It is sometimes possible to examine mass transfer at a microscopic level to obtain expressions for the rate, and this is traditionally done in later courses covering mass transfer, but we shall not follow that path here. Instead, as with reacting systems, we will construct the simplest form possible that is compatible with experimental observation.

It is commonly observed that, if we wait long enough, equilibrium is reached in a batch system and there is a fixed functional relationship between the concentrations of the solute in the two phases. That is, at equilibrium,

$$c_{Ae}^I = f(c_{Ae}^{II}),$$ (10.5)

where the subscript e denotes equilibrium. There is a large amount of such data in the technical literature. Figure 10.2 shows a typical set of experimental equilibrium data, in this case for the concentration of acetone in water and 1,1,2-trichloroethane. The data were obtained by adding a known amount of acetone to the two solvents, after which the system was thoroughly mixed and permitted to come to equilibrium. The two liquid phases were then separated and the concentration of acetone present in each phase was measured. The concentration of acetone in the organic phase is a unique function of the concentration in the aqueous phase; the functional relation is linear at low concentrations, but becomes nonlinear at higher concentrations.

The time derivatives in Equations 10.3 and 10.4 must equal zero at equilibrium, indicating that $a\mathbf{r}_A$ must go to zero. Since a cannot go to zero, it follows that $\mathbf{r}_A = 0$ at equilibrium. The functional form of the rate must be such that its vanishing at equilibrium is compatible with Equation 10.5. We also have another important piece of information about the rate. Suppose that c_A^I is greater than its equilibrium value for a given c_A^{II}; in that case, there must be a net transfer of species A from Phase I (the continuous phase) to Phase II (the dispersed phase), so $\mathbf{r}_{A+}^{II} > \mathbf{r}_{A-}^{II}$ and, from Equation 10.2, $\mathbf{r}_A > 0$. Thus,

$$c_A^I - f(c_A^{II}) > 0 \Rightarrow \mathbf{r}_A > 0. \tag{10.6a}$$

Similarly,

$$c_A^I - f(c_A^{II}) < 0 \Rightarrow \mathbf{r}_A < 0. \tag{10.6b}$$

The *simplest* form compatible with Equations 10.6a-b is

$$\mathbf{r}_A = K_m[c_A^I - f(c_A^{II})], \tag{10.7}$$

where the *overall mass transfer coefficient* K_m has dimensions of length/time. (Note that Equation 10.7 is, in fact, a *definition* of the mass transfer coefficient, since the rate and concentrations can, in principle, be measured independently.)

The equilibrium relation, Equation 10.5, is often written

$$c_{Ae}^I = Mc_{Ae}^{II}, \tag{10.8}$$

where the *distribution coefficient M* will not be a constant in general, except at low concentrations (*cf.* Figure 10.2).* In that case, the rate expression will be written

$$\mathbf{r}_A = K_m[c_A^I - Mc_A^{II}]. \tag{10.9}$$

This form is completely general as long as K_m and M are both allowed to be functions of concentration. For simplicity of illustration only we will take the coefficients to be constants in our treatment. It is sometimes helpful to note that Equation 10.9 has the form of a *driving force*: the distance from equilibrium, $c_A^I - Mc_A^{II}$, divided by a *resistance*, K_m^{-1}, much like Ohm's law in electricity. It is shown in texts on mass transfer that K_m can be expressed in terms of resistances in series, in analogy to the corresponding relation in electricity,

$$\frac{1}{K_m} = \frac{1}{k^I} + \frac{M}{k^{II}},$$

where k^I and k^{II} are associated with transport in the interfacial regions of the individual phases.

---

* Equation 10.8 is known as *Nerst's Law* when M is a constant. This is one of many "laws" in physical chemistry and physics that carry Nerst's name. It is not a law of nature, of course.

10.3.3 A Rate Experiment

Batch mass transfer experiments to measure K_m are usually difficult to carry out for two reasons: First, equilibrium is typically achieved very rapidly, so reasonable transient data are difficult to obtain. Second, the product aK_m is the quantity that appears in the equations, so independent knowledge of the interfacial area a is necessary in order to extract K_m from experimental data, but the area is usually unknown and may, in fact, be changing during the course of the experiment. We have designed an experiment that occurs over several minutes in which the interfacial area is known, thus permitting direct measurement of the interfacial mass transfer coefficient. We dissolve tablets of table salt (NaCl) in distilled water, and we follow the dissolved salt concentration using a conductivity probe, as we did in the experiment leading to Figure 4.4.

The mathematical description of the salt-water system is a bit simpler than the general formulation because the salt phase is a pure material, so $c_A^{II} = \rho^{II}$ (remember, we are using mass units) and Equations 10.3II and 10.4II are identical. Furthermore, the equilibrium concentration of dissolved salt is simply the *saturation* concentration, which we denote c_{As}^I; c_{As}^I is a function only of temperature. Hence, we replace Mc_A^{II} in the rate expression with c_{As}^I and write

$$\mathbf{r}_A = K_m[c_A^I - c_{As}^I]. \tag{10.10}$$

Equations 10.3 and 10.4 thus become

$$\frac{d\rho^I V^I}{dt} = -a\mathbf{r}_A = -K_m a[c_A^I - c_{As}^I], \tag{10.11I}$$

$$\frac{d\rho^{II} V^{II}}{dt} = a\mathbf{r}_A = K_m a[c_A^I - c_{As}^I], \tag{10.11II}$$

$$\frac{dc_A^I V^I}{dt} = -a\mathbf{r}_A = -K_m a[c_A^I - c_{As}^I]. \tag{10.12}$$

The basic data for the experiment are shown in Table 10.1, and they lead to some further simplifications. The total mass of the salt is one-third of one percent of the mass of the water. Thus, there will never be a significant change in the density or volume of the aqueous phase, so we can disregard the overall mass balance for Phase I (Equation 10.11I) and take V^I as a constant throughout the experiment. Furthermore, c_A^I is more than two orders of magnitude less than c_{As}^I at all times, so $c_{As}^I - c_A^I \sim c_{As}^I$. Also, ρ^{II} is a constant. Thus, taking the constant terms outside the derivatives, the experiment can be described by two equations,

$$\rho^{II} \frac{dV^{II}}{dt} = -K_m a c_{As}^I, \tag{10.13a}$$

$$V^I \frac{dc_A^I}{dt} = K_m a c_{As}^I. \tag{10.13b}$$

Table 10.1. *Experimental conditions and concentration of dissolved salt as a function of time.*

Number of tablets	$N = 30$
Density of salt	$\rho^{II} = 2.16 \text{ g/cm}^3$
Total mass of tablets	19.2 g
Initial volume of tablets	$V_0^{II} = 19.2/2.16 = 8.85 \text{ cm}^3$
Volume of water	$V^I = 6{,}000 \text{ cm}^3$
Surface-to-volume factor	$\alpha = 5.32$
Saturation concentration	$c_{As}^I = 0.360 \text{ g/cm}^3$

Time (seconds)	$1000 \, c_A^I \ (\text{g/cm}^3)$
0	0
15	0.30
30	0.35
45	0.64
60	0.89
75	1.08
90	1.10
105	1.24
120	1.40
135	1.49
150	1.68
165	1.76
195	2.06
200	2.14
240	2.31
270	2.43

Equations 10.13a and 10.13b can be added to give

$$\rho^{II}\frac{dV^{II}}{dt} + V^I\frac{dc_A^I}{dt} = 0,$$

which, on integration, yields

$$\rho^{II}V_0^{II} - \rho^{II}V^{II} = V^I c_A^I, \tag{10.14}$$

where V_0^{II} is the initial volume of solid. $\rho^{II}V_0^{II}$ is, of course, the initial mass of salt.

The interfacial area a is related to the volume V^{II} through the solid geometry of the tablets. We assume that the tablets retain their shape as they dissolve, which turns out to be a good assumption until the very end of the experiment, when the tablets begin to crumble. The N tablets are assumed to be identical, so the volume per tablet is V^{II}/N. The interfacial area per tablet is then $\alpha[V^{II}/N]^{2/3}$, where α is a constant that depends on tablet geometry. For a sphere, $\alpha = [36\pi]^{1/3} = 4.84$; for a cube, $\alpha = 6$; and for a square cylinder (height = diameter), $\alpha = 5.50$. The tablets used in this experiment were nearly square cylinders, with $\alpha = 5.32$. The total surface area is then N times the area per tablet,

$$a = N\alpha[V^{II}/N]^{2/3} = \alpha N^{1/3}[V^{II}]^{2/3}. \tag{10.15}$$

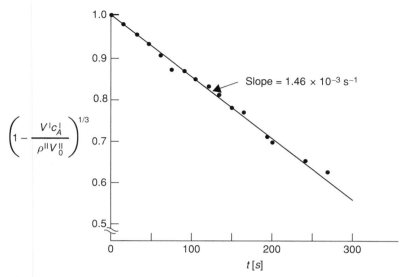

Figure 10.3. Rate of solution of NaCl in water. The data are plotted according to Eq. 10.18; $1 - V^I c_A^I / \rho^{II} V_0^{II}$ is the fraction of NaCl that is undissolved.

Equation 10.13a then becomes

$$\rho^{II} \frac{dV^{II}}{dt} = -K_m \alpha N^{1/3} c_{As}^I [V^{II}]^{2/3}, \tag{10.16a}$$

which is a separable equation that can be rearranged for quadrature as

$$\frac{dV^{II}}{[V^{II}]^{2/3}} = [-K_m \alpha N^{1/3} c_{As}^I / \rho^{II}] dt. \tag{10.16b}$$

Upon integration we obtain

$$V^{II}(t) = \left\{ [V_0^{II}]^{1/3} - \frac{K_m \alpha N^{1/3} c_{As}^I t}{3\rho^{II}} \right\}^3. \tag{10.17}$$

Finally, we substitute Equation 10.17 into Equation 10.14 and rearrange to obtain a form that is convenient for comparison with data:

$$\left(1 - \frac{V^I c_A^I}{\rho^{II} V_0^{II}} \right)^{1/3} = 1 - \frac{K_m \alpha N^{1/3} c_{As}^I}{3[V_0^{II}]^{1/3} \rho^{II}} t. \tag{10.18}$$

Note that $V^I c_A^I / \rho^{II} V_0^{II}$ is simply the fraction of the total salt that is in solution, and $1 - V^I c_A^I / \rho^{II} V_0^{II}$ is the fraction that is undissolved.

The data are plotted according to Equation 10.18 in Figure 10.3. The data do follow a straight line passing through unity, as required. Using the experimental value 1.46×10^{-3} s^{-1} of the slope and the data given in Table 10.1, we then compute $K_m = 3.3 \times 10^{-3}$ cm/s. We shall not show any further data, but it is found that K_m in two-phase tank-type systems is almost always within an order of magnitude of the value found here. Since the range of the interfacial mass transfer coefficient is so limited (compare the range for reaction rates in Chapter 6), the significant problem is determination of the interfacial area for each situation. The interfacial area varies greatly depending on the degree of agitation, and there have been many studies to develop relationships between power to the mixer, geometry, physical properties, and interfacial area.

10.3.4 Approach to Equilibrium

We remarked previously that batch experiments for mass transfer data are often too rapid to obtain useful data. This statement can be quantified now that we have a reasonable estimate of the magnitude of K_m. We will assume that we have two liquid phases that are initially of equal volume, and that the total amount of dissolved substance A that is transferred between the phases is so small that changes in volume are negligible. The overall mass balance equations, Equations 10.3I and II, therefore provide no useful information. Equations 10.4I and II for the species balance in the two phases then become

$$V\frac{dc_A^I}{dt} = -a\mathbf{r}_A = -K_m a[c_A^I - Mc_A^{II}], \tag{10.19I}$$

$$V\frac{dc_A^{II}}{dt} = +a\mathbf{r}_A = +K_m a[c_A^I - Mc_A^{II}]. \tag{10.19II}$$

We have not included superscripts I or II for the volumes since they are equal and constant. M will be taken as constant.

Addition of the Equations 10.19I and II gives $V\dfrac{dc_A^I}{dt} + V\dfrac{dc_A^{II}}{dt} = 0$, or

$$c_A^I - c_{A0}^I + c_A^{II} - c_{A0}^{II} = 0. \tag{10.20}$$

c_{A0}^I and c_{A0}^{II} are the values of c_A^I and c_A^{II}, respectively, at $t = 0$. Substitution for c_A^{II} in Equation 10.19I then gives

$$\frac{dc_A^I}{dt} = -\frac{K_m a(1 + M)}{V}c_A^I + \frac{\left(c_{A0}^I + c_{A0}^{II}\right)MK_m a}{V}. \tag{10.21}$$

This is an equation of a form that we have solved many times before, and the solution can be written

$$c_A^I(t) = c_{Ae}^I\left\{1 - \left[1 - \frac{c_{A0}^I}{c_{Ae}^I}\right]e^{-[K_m a(1+M)t/V]}\right\}, \tag{10.22}$$

where the equilibrium concentration c_{Ae}^I is computed from the equilibrium relation and Equation 10.20 as

$$c_{Ae}^I = \frac{\left(c_{A0}^I + c_{A0}^{II}\right)M}{1 + M}. \tag{10.23}$$

According to Equation 10.22, the system will be at equilibrium when the exponential term is negligible. This will occur for an exponent of about −3 ($e^{-3} \sim 0.05$), so the total time t_e for the experiment will be determined approximately by

$$\frac{K_m a(1 + M)t_e}{V} \approx 3. \tag{10.24}$$

We will take $M \sim 2$, based on the data in Figure 10.1, and we have seen that 3×10^{-3} cm/s is a reasonable value for K_m. Thus, $t_e \sim 10^3 V/3a$, where lengths are in centimeters and time is in seconds. Now, if the agitation is such that that one phase is effectively dispersed into N droplets then, according to Equation 10.15, $a = \alpha N^{1/3} V^{2/3}$, where the shape factor α is approximately 5. Thus, $t_e \approx 10^2 (V/N)^{1/3}$.

V/N is the volume of a typical droplet. Droplets with a diameter of about 1mm are easily obtained in most low-viscosity systems, so $V/N \sim (1/6)\pi(10^{-1})^3 \sim 5 \times 10^{-4}$, and $t_e \sim 10$ s. Thus, the system will be more than halfway to equilibrium after three seconds and completely there by ten seconds. This is to be compared to timescales of order minutes or even hours for aqueous reacting systems, as shown in Chapter 6.

The order-of-magnitude analysis carried out in this section is important in its own right, because it provides useful insight into the mechanics of interfacial mass transfer that we will exploit subsequently. It is also important, however, because it represents the type of "back-of-the-envelope" calculation that engineers are routinely expected to carry out. Good engineering requires the ability to obtain quick estimates of the relative importance of the various physical phenomena that are occurring in any process, and rough calculations like this one exemplify the way in which that procedure is carried out.

10.3.5 Further Comments

It is important to note the essential difference in usefulness between the batch experiment for a mass transfer system and the batch reactor experiments described in Chapter 6. We have already observed that the batch mass transfer experiment may be more difficult to carry out than the reaction experiment. This is true not only because of the rapid approach to equilibrium, but also because the presence of two phases in intimate contact can often increase the sampling and measurement problems considerably. There is a second extremely important difference that might not be obvious from the discussion thus far. The data from the single-phase batch reactor experiments are sufficient to enable us to compute the design specifications for a continuous processing unit. Such is not the case with the two-phase system. Both the rate and the interfacial area are required, and it is rare that data on interfacial area obtained in a small-scale batch system provide meaningful information about interfacial areas in large, continuous-flow devices.

10.4 Continuous-Flow Two-Phase Systems

Continuous-flow two-phase systems are ones in which both phases are continuously fed to and removed from the system. They are widely employed for the various unit operations described in Section 10.2 and may be carried out in tank-type or tubular geometries. They are also used in the laboratory to collect experimental data. (Semicontinuous systems, in which one phase is stationary and only one phase flows through the system, are also used, but we will not address them here.) Our

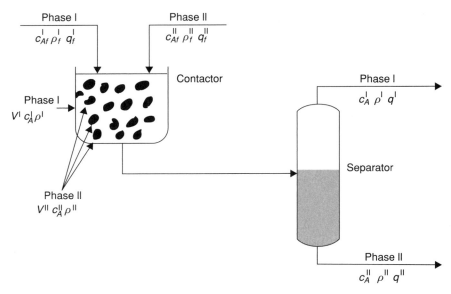

Figure 10.4. Schematic of a continuous-flow two-phase process.

goal here is to gain an appreciation of the issues in two-phase design and operation by examining an elementary well-stirred continuous-flow tank-type mass transfer system in detail.

A schematic of the continuous-flow system is shown in Figure 10.4. Two pieces of equipment are shown, a tank-type contactor and a device that effects a separation between the phases. We assume that the contents of the contactor are well mixed, so that the two phases are in intimate contact and there is no spatial variation of the concentration of a species in either phase. We also assume that all mass transfer takes place in the contactor and that the sole function of the separator is to separate the two phases that are mixed in the contactor. We do not consider the separator operation in detail; it will often be no more than a holding tank large enough to allow separation by gravity. (In some gas-liquid systems the separation takes place in the contactor, and a distinct separation device is not needed.) The concentrations in the exit stream from the separator are assumed to be the same as those that we would find if we were to sample the phases in the contactor. This is a reasonable assumption since, without agitation in the separator, there is little interfacial area available for mass transfer. The contactor and separator together are frequently referred to as a *stage.*

Our control volumes will be the same as those designated for the batch two-phase systems, and we can develop the model equations by applying conservation of mass exactly as we did in Section 10.3. The overall mass balance equations are

$$\frac{d\rho^I V^I}{dt} = q_f^I \rho_f^I - q^I \rho^I - a\mathbf{r}_A, \tag{10.25I}$$

$$\frac{d\rho^{II} V^{II}}{dt} = q_f^{II} \rho_f^{II} - q^{II} \rho^{II} + a\mathbf{r}_A. \tag{10.25II}$$

Note that we do not consider any mass transfer in the separator, so the streams issuing from the separator have the same compositions as those in the contactor. This is conceptually the same as if we had assumed that separated streams issue directly from the contactor.

The component equations will be written for the case in which a single species is being transferred. This is a common situation and allows us to develop the important concepts without the algebra becoming too complex. Concentrations everywhere in the tank and at the separator exit are denoted by c_A^I and c_A^{II}:

$$\frac{dc_A^I V^I}{dt} = q_f^I c_{Af}^I - q^I c_A^I - a\mathbf{r}_A, \tag{10.26I}$$

$$\frac{dc_A^{II} V^{II}}{dt} = q_f^{II} c_{Af}^{II} - q^{II} c_A^{II} + a\mathbf{r}_A. \tag{10.26II}$$

Transient behavior is important for startup, shutdown, and control of the system, but the basic design is carried out for the steady state, when time derivatives are zero. We thus write, using the rate expression Equation 10.9,

$$0 = q_f^I \rho_f^I - q^I \rho^I - K_m a [c_A^I - M c_A^{II}], \tag{10.27I}$$

$$0 = q_f^{II} \rho_f^{II} - q^{II} \rho^{II} + K_m a [c_A^I - M c_A^{II}], \tag{10.27II}$$

$$0 = q_f^I c_{Af}^I - q^I c_A^I - K_m a [c_A^I - M c_A^{II}], \tag{10.28I}$$

$$0 = q_f^{II} c_{Af}^{II} - q^{II} c_A^{II} + K_m a [c_A^I - M c_A^{II}]. \tag{10.28II}$$

A complete solution to this set of equations will require the constitutive equations between densities and phase compositions.

The general problem with composition-dependent densities is sometimes important, but we can accomplish our goals by dealing with the more limited case in which the total amount of A that is transferred between phases is insufficient to have an effect on the phase volumes. In that case, $q_f^I = q^I$, $q_f^{II} = q^{II}$, and only the component equations 10.28I and 10.28II are needed to describe the system fully. This approximation leads to negligible error for most applications. Combining Equations 10.28I and 10.28II leads to an alternate equation relating the two concentrations:

$$c_A^I = c_{Af}^I + \frac{q^{II}}{q^I} \left[c_{Af}^{II} - c_A^{II} \right]. \tag{10.29}$$

Note that Equation 10.29 does not include any terms involving the mass transfer rate; it is simply an overall mass balance that equates the total mass flow rate of A into the system, including both phases, to the total mass flow rate of A out. Any two of the three equations 10.28I, 10.28II, and 10.29 are independent and can be used to analyze the process. There are eight quantities in these equations: $q^I, q^{II}, c_{Af}^I, c_{Af}^{II}, c_A^I, c_A^{II}, M$, and $K_m a$; six independent quantities must be specified, and the two independent equations can then be used to find the other two. (K_m and a always appear as a product, so these two experimental quantities cannot be determined independently as outputs from any experiment that is analyzed by this set of equations.)

10.4.1 Equilibrium Stage

A typical problem is to determine the effluent concentrations. Equations 10.28II and 10.29 can be combined to obtain the following result for c_A^{II}:

$$c_A^{II} = \frac{c_{Af}^I}{\lambda + M + (q^{II}/K_m a)} + \frac{c_{Af}^{II}}{1 + \frac{M}{\lambda + (q^{II}/K_m a)}}. \tag{10.30}$$

Here, $\lambda = q^{II}/q^I$. Note that the volume, or, equivalently, the holdup in the tank, does not appear for either phase. The important transport quantity is the ratio of the flow rate of the dispersed phase to the rate of interfacial mass transfer.

The ratio $q^{II}/K_m a$ can be written

$$\frac{q^{II}}{K_m a} = \frac{V^{II}/K_m a}{V^{II}/q^{II}} = \frac{V^{II}/K_m a}{\theta^{II}}, \tag{10.31}$$

where θ^{II} is the residence time of the dispersed phase. The calculations in Section 10.3.4 show that $V^{II}/K_m a$ is of the order of one minute if the agitation is sufficient to produce droplets with diameters of the order of 1 mm, and M is of order unity. If the system is designed for a holdup of, say, five to ten minutes, so that $q^{II}/K_m a \ll 1$, then the $q^{II}/K_m a$ terms in Equation 10.31 can be neglected (they can be neglected in the second term only if $\lambda = q^{II}/q^I$ is of order unity) and the effluent concentration can be written

$$c_A^{II} = \frac{c_{Af}^I + \lambda c_{Af}^{II}}{\lambda + M}. \tag{10.32}$$

The rate of mass transfer does not appear, and the calculations can be carried out without knowledge of $K_m a$! Clearly this is an extraordinary simplification.

Equation 10.32 is formally equivalent to taking the limit $K_m a \to \infty$ in Equation 10.30. The result is known as an *equilibrium stage* and can be obtained, with considerably less physical insight, in a more direct manner. We simply take Equation 10.29, which is always valid, and assume that we have equilibrium, in which case we write $c_A^I = M c_A^{II}$, from which Equation 10.32 follows directly. This is the reason for the name *equilibrium stage*. The concept of the equilibrium stage is used extensively in design calculations for the unit operations touched on in Section 10.2. We illustrate the concept with a few examples here, and we will return to this important topic in Chapter 11.

EXAMPLE 10.1 An aqueous solution containing 200 g/L of acetone is to be purified by continuous extraction with pure trichloroethane. How much acetone can be removed if the flow of both aqueous and organic streams is 10 L/min?

We assume for these calculations that water and trichloroethane are insoluble. This is a reasonable assumption for our purposes here; the National Institute of Standards and Technology (NIST) database[*] summarizes five studies and concludes that 0.0012 grams of trichloroethane or less will dissolve in 1 gram of water in the range $0 - 50°C$. (At a level of over 1,000 ppm the water would probably have to be purified before it could be discharged. More likely, it would be recycled for process use, where the very small amount of trichloroethane would be unlikely to be relevant.) We assume that the contactor is an equilibrium stage, the flow rates are constant, and the solvent feed contains pure trichloroethane ($c_{Af}^I = 0$). Because the flow rates are equal, $\lambda = 1$. From Figure 10.1 we have $M = 2$. Hence, from Equation 10.33,

$$c_A^{II} = \frac{\lambda c_{Af}^{II}}{\lambda + M} = \frac{1 \times 200}{1 + 2} = 66.7 \text{ g/L}, \quad c_A^I = M c_A^{II} = 2 \times 66.7 = 133.3 \text{ g/L}.$$

That is, the acetone content in the aqueous phase is reduced from 200 to 66.7 g/L in this equilibrium stage.

EXAMPLE 10.2 Suppose now that the flow of organic in Example 10.1 is increased from 10 to 20 L/min and the flow rate of the aqueous phase is maintained at 10 L/min.

We now have $\lambda = q^{II}/q^I = 0.5$, and

$$c_A^{II} = \frac{\lambda c_{Af}^{II}}{\lambda + M} = \frac{0.5 \times 200}{0.5 + 2} = 40 \text{ g/L}, \quad c_A^I = M c_A^{II} = 2 \times 40 = 80 \text{ g/L}.$$

By doubling the amount of solvent the acetone content in the aqueous stream is reduced only from 66.7 to 40 g/L.

EXAMPLE 10.3 As an alternative to the increased organic flow in Example 10.2, suppose that the aqueous effluent in Example 10.1 is taken to a second stage and again contacted with a pure trichloroethane stream with a flow rate of 10 L/min.

For the second stage, $c_{Af}^{II} = 66.7$ g/L and $\lambda = 1$. Thus,

$$c_A^{II} = \frac{\lambda c_{Af}^{II}}{\lambda + M} = \frac{1 \times 66.7}{1 + 2} = 22.2 \text{ g/L}, \quad c_A^I = M c_A^{II} = 2 \times 22.2 = 44.4 \text{ g/L}.$$

Thus, by using the same amount of solvent as in Example 10.2, but by dividing the total between two consecutive contacting stages, the residual acetone in the water stream is reduced by nearly a factor of two. We saw a similar result in the treatment of cross-flow dialysis in Section 5.3.

10.4.2 Deviation from Equilibrium

Energy in the form of agitation must be put into two-phase systems to generate adequate interfacial area for efficient mass transfer. Hence, it is evident that there will be situations where equilibrium cannot be attained in a stage. This has led to

[*] See the Bibliographical Notes at the end of the chapter.

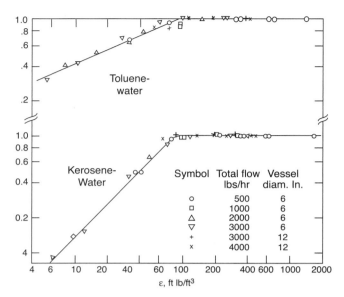

Figure 10.5. Residual efficiency in a stirred tank as a function of energy input/unit volume. Data of A. W. Flynn and R. E. Treybal, *AIChE J.*, **1**, 324–328 (1955). Reproduced with the permission of the American Institute of Chemical Engineers.

attempts to correlate $K_m a$ with quantities like the power input per unit volume in order to compute effluent composition. One quantity that is frequently used in unit operations is the *stage efficiency,* defined as

$$\mathbf{E}_f = \frac{c_A^{II} - c_{Af}^{II}}{c_{Ae}^{II} - c_{Af}^{II}}. \tag{10.33}$$

The stage efficiency defined in this manner is nothing more than the fractional approach to equilibrium. Noting that $c_{Ae}^{II} = c_A^{I}/M$ and substituting Equation 10.33 into Equation 10.29, we obtain an equation for the effluent in terms of this additional parameter:

$$c_A^{II} = \mathbf{E}_f \frac{c_{Af}^{I} + \lambda c_{Af}^{II}}{\lambda \mathbf{E}_f + M} + \frac{M(1 - \mathbf{E}_f)c_{Af}^{II}}{\lambda \mathbf{E}_f + M}. \tag{10.34}$$

When $\mathbf{E}_f \to 1$, Equation 10.34 reduces to Equation 10.32 for the equilibrium stage. As might be expected, the efficiency is simply related to the mass transfer coefficient. Through comparison of Equations 10.31 and 10.35 the relationship can be established as

$$\mathbf{E}_f = \frac{M}{M + q^{II}/K_m a}. \tag{10.35}$$

The stage efficiency can be correlated with design variables. Figure 10.5, for example, shows some data of Flynn and Treybal for interfacial transfer of benzoic acid in toluene-water and kerosene-water systems. The *residual efficiency* E_R plotted in the figure is defined by

$$E_R = \frac{E_f - E_{f0}}{1 - E_{f0}}. \tag{10.36}$$

E_{f0} is the measured efficiency at zero agitator speed (which is itself high). Clearly, as $E_R \to 1$, $E_f \to 1$. ε is the energy per unit volume,

$$\varepsilon = \frac{P}{q^I + q^{II}}, \tag{10.37}$$

where P is the power supplied to the agitator. The correlation appears to work for each system for the type of mixer used (six-bladed turbine impellors), and it is evident that beyond an energy input of approximately 100 ft lb_f/ft^3 (or about 5 kJ/m^3) an equilibrium stage can be assumed.

Many other correlations for the efficiency and the quantity $K_m a$ are available in the published literature. A further discussion of this important practical topic, which is typically covered in a subsequent course in separations or unit operations, is beyond the scope of our introductory treatment, however.

10.5 Two-Phase Reactors

Many important chemical reactions take place in two-phase reactors, in which a reactant is transferred from the phase in which it is fed to the reactor to a second phase in which it reacts. This is often done to allow the reactants to come into contact with one another uniformly throughout the reactor when high localized concentrations near the entry might lead to undesirable side reactions and, perhaps, fouling or plugging. In other cases, one of the reactants is naturally present in another phase, as in the bio-oxidation of liquid waste discussed in Chapter 8, where oxygen must be transferred from an air stream to solution in the liquid.

We will consider here only the simplest case of a two-phase reactor. Phase I contains a species, A, that when transferred to Phase II undergoes the irreversible first-order or pseudo-first-order reaction A $\to$ products. We will assume that the total amount of A transferred between the phases is sufficiently small that flow rate changes because of the mass transfer and reaction are negligible, so the overall mass balances can be neglected. The flow rates are steady, so feed and effluent flow rates are equal and the phase volumes are constant. The component mass balances are then

$$V^I \frac{dc_A^I}{dt} = q^I c_{Af}^I - q^I c_A^I - a\mathbf{r}_A, \tag{10.38I}$$

$$V^{II} \frac{dc_A^{II}}{dt} = -q^{II} c_A^{II} + a\mathbf{r}_A - V^{II} r_{A-}. \tag{10.38II}$$

$\mathbf{r}_A$ is as given previously in Equation 10.9, and for the pseudo-first-order reaction $r_{A-} = kc_A^{II}.$* At steady state the time derivatives are zero, and, after substituting the rate expressions, the resulting equations are easily solved for c_A^{II}:

$$c_A^{II} = \frac{c_{Af}^I}{M + [1 + k\theta^{II}][\lambda + q^{II}/K_m a]}. \tag{10.39}$$

Here, $\lambda = q^{II}/q^I$ and $\theta^{II} = V^{II}/q^{II}$.

Some interesting limiting cases follow. If agitation is sufficient to ensure that $q^{II}/K_m a \ll \lambda$, or equivalently, $q^I/K_m a \ll 1$, then the mass transfer term can be neglected and we obtain

$$q^I/K_m a \ll 1 : c_A^{II} \to \frac{c_{Af}^I}{M + \lambda[1 + k\theta^{II}]}. \tag{10.40}$$

The physical meaning of this limit is that reactant is transferred from Phase I to Phase II much more rapidly than it is carried out in Phase I by flow. This case is often referred to as *reaction limited*, since the mass transfer is sufficiently rapid that the reaction rate is the only term of importance. This is analogous to the equilibrium stage and reduces to it as $k\theta^{II} \to 0$.

The second limit of interest, which can occur in a highly viscous system where efficient agitation is difficult, is one in which a is sufficiently small that $K_m a/q^{II}$ is small compared to λ, or, equivalently, $q^I/K_m a \gg 1$. In that case we would also have $q^{II}/K_m a \gg M$ and both the λ and M terms will drop out of Equation 10.40, leading to

$$q^I/K_m a \gg 1 : c_A^{II} \to \frac{K_m a}{q^{II}} \frac{c_{Af}^I}{1 + k\theta^{II}}. \tag{10.41}$$

The physical meaning of this limit, which is often referred to as *mass transfer limited*, is that reactant is carried out in Phase I by flow more rapidly than it is transferred to Phase II. A characteristic of such a system, which is often encountered in the production of polymers, is that the conversion depends strongly on the design and intensity of the mixing device even when complete mixing in the sense of spatial uniformity has been achieved.

The concepts of reaction limited and mass transfer limited multiphase reacting systems are of considerable importance in applications beyond the simple tank-type device considered here. The recognition of such limits guides design considerations. There is little point, for example, in expending the energy cost to move to a transport regime that is considerably more efficient than the transition region between mass transfer and reaction limitation ($q^I/K_m a \sim 1$ in the context of the discussion here). Catalysts are typically constructed to operate just beyond the mass transfer limit. This concept also has a strong evolutionary basis in biological systems, where most

---

* The reaction rate must now be interpreted in mass/time rather than in moles/time. It is straightforward to show that the first-order rate constant k will be the same with either system of units because of the linearity and homogeneity of the rate expression. This would not be the case in general, and species molecular weights would have to be known to transform from molar to mass rates.

reacting systems involve interfacial mass transfer; it is found that most biological systems operate just a bit more efficiently than the mass transfer limit.

10.6 Concluding Remarks

In reviewing this chapter it is helpful to return once more to the logic diagram, Figure 4.2. Note the critical role played by the selection of the control volume and the parallel between the development of the rate of mass transfer here and the rate of reaction in Chapter 6.

The estimate of the speed at which phase equilibrium is attained is of particular importance and should be examined carefully. Calculations of this type often lead to substantial simplifications in engineering problems; in the case of mass transfer, the consequence is the equilibrium stage. Nearly all designs for staged separation systems are based on the equilibrium stage, sometimes taking the efficiency into account. The design of liquid-liquid separation systems often requires taking the mutual solubility of the solvents into account, and this is addressed in the specialized courses in the curriculum and the relevant textbooks.

Bibliographical Notes

This chapter closely follows Chapter 8 in

Russell, T. W. F., and M. M. Denn, *Introduction to Chemical Engineering Analysis*, Wiley, New York, 1972.

There is very similar material, based on the same source, in

Russell, T. W. F., A. S. Robinson, and N. J. Wagner, *Mass and Heat Transfer: Analysis of Mass Contactors and Heat Exchangers*, Cambridge, New York, 2008.

Two-phase separation systems are discussed in all editions of the *Chemical Engineers' Handbook,* now published as

Green, D., and R. Perry, Eds., *Perry's Chemical Engineers' Handbook*, 8th Ed., McGraw-Hill, New York, 2007.

This edition is available electronically. The topic is also addressed in books on unit operations, such as

Foust, A. S., L. A. Wenzel, C. W. Clump, L. Maus, and L. B. Andersen, *Principles of Unit Operations*, 2nd Ed., Wiley, New York, 1980; reprinted by Robert Krieger, Miami, 1990.

McCabe, W. L., J. C. Smith, and P. Harriott, *Unit Operations of Chemical Engineering*, 7th Ed, McGraw-Hill, New York, 2004.

Foust and coworkers was published three decades ago, but the material relevant to this chapter is still appropriate.

The determination of mass transfer rates and interfacial areas is addressed in most books on unit operations and books on mass transfer, as well as books with titles that include the words *transport phenomena, transfer processes,* or *momentum, mass,* and *heat transfer,* but rarely in books in which the title contains only the words *heat* and *mass transfer,* where the mass transfer typically gets short shrift. (The book by Russell and coworkers cited previously is an important exception to this rule.)

Phase equilibrium data can be found in handbooks and in the NIST Physical Properties database. All NIST physical property databases were available online at http://srdata.nist.gov/gateway/ at the time of writing. The URL for the solubility database is http://srdata.nist.gov/solubility/.

There is a discussion of the transition between reaction and mass transfer controlled regimes that includes biological systems in

> Weisz, P. B., "Diffusion and reaction: An interdisciplinary excursion," *Science,* **179**, 433–440 (1973).

PROBLEMS

10.1. Derive Equation 10.29 by carrying out a mass balance for species A on a control volume that encompasses both Phase I and Phase II. Then assume that A is in equilibrium between the two phases and derive Equation 10.32 for the equilibrium stage.

10.2. One step in the preparation of nicotine sulfate from tobacco is the extraction of nicotine from aqueous solution by kerosene. Table 10.P1 shows equilibrium data for the distribution of nicotine between water and kerosene. Compute the kerosene-water ratio required to remove 90 percent of the nicotine from the aqueous stream in a single equilibrium stage.

Table 10.P1. *Equilibrium distribution of nicotine between aqueous and organic phases. Data of J. B. Claffey, C. O. Badgett, J. J. Skalamera, and G. W. M. Phillips,* Industrial & Engineering Chem., *92, 166–171 (1950).*

Nicotine in aqueous phase (I) (g/L)	Nicotine in organic phase (II) (g/L)
0.62	0.39
1.49	0.96
2.92	2.07
5.70	4.18
11.8	8.22
17.9	12.2
24.9	15.3
31.5	18.9

10.3. The data in Table 10.P2 are from a batch experiment measuring the distribution of octanoic acid between an aqueous phase consisting of a solution of corn syrup in water and an organic xylene phase. The aqueous phase was continuous, and the concentration in the aqueous phase was measured with a calibrated conductivity probe. Initially, 2 L of the aqueous phase was placed in a tank with 0.2 L of xylene and agitated until the drop size of the dispersed organic phase had equilibrated. 0.25 L of an aqueous phase containing octanoic acid was then added and the concentration in the continuous phase was recorded as a function of time. Determine $K_m a$ and compare to the value of 75 cm³/s reported by the authors of the original article.

Table 10.P2. *Concentration of octanoic acid in aqueous phase. Data of J. H. Rushton, S. Nagata, and T. B. Rooney, AIChE J., 10, 298–302 (1964).*

t (s)	$c_A{}^I \times 10^4$ (g-mol octanoic acid/L)
0	2.75
10	2.13
20	1.72
30	1.45
40	1.23
60	1.03
80	0.94
120	0.83
∞	0.78

10.4. A useful model of interfacial mass transfer makes three assumptions: (i) there is a difference between the concentrations of A at the interface, denoted c_A^{I*} and c_A^{II*} in Phases I and II, respectively, and the bulk concentrations c_A^I and c_A^{II}; (ii) the rates of transfer in each phase are proportional to the differences between the bulk and interfacial concentrations: $\mathbf{r}_{A+}^I - \mathbf{r}_{A-}^I = k^I \left(c_A^{I*} - c_A^I \right)$ and $\mathbf{r}_{A+}^{II} - \mathbf{r}_{A-}^{II} = k^{II} \left(c_A^{II*} - c_A^{II} \right)$; and (iii) the interfacial concentrations are in equilibrium: $c_A^{I*} = M c_A^{II*}$. Show that Equation 10.9 follows directly from this formulation, with $\frac{1}{K_m} = \frac{1}{k^I} + \frac{M}{k^{II}}$.

10.5. Trace amounts of phenol are to be extracted from an aqueous stream (Phase I) using pure xylene as a solvent. From data in the *Chemical Engineers' Handbook* we have $M = 1.4$.

    a. What xylene/water flow ratio is required to remove 90 percent of the phenol in a single equilibrium stage?

    b. Suppose two equilibrium stages are to be used, using the aqueous effluent from Stage 1 as a feed to Stage 2. What xylene/water flow ratio is required for each stage?

10.6. Data for the distribution of picric acid in a benzene-water system at 15°C are shown in Table 10.P3. (Note that the concentrations are in molar units, and that Nerst's "law" clearly does not apply to these data.) You wish to employ an equilibrium stage to extract picric acid from a water stream with a picric acid concentration of 0.02 g-mol/L using a pure benzene stream.[*] Assume $q^I = q^{II} = 50$ L/min. What fraction of the picric acid is removed? (Hint: The most efficient approach to solving this problem is to develop a graphical scheme.)

Table 10.P3. *Distribution of picric acid between water and benzene. Data from F. Daniels,* Outlines of Physical Chemistry, *John Wiley, New York, 1948.*

Picric acid in benzene, c_A^I (g-mol/L)	Picric acid in water, c_A^{II} (g-mol/L)
0.000932	0.00208
0.00225	0.00327
0.0101	0.00701
0.0199	0.0101
0.0500	0.0160
0.100	0.0240
0.180	0.0336

Table 10.P4. *Efficiency as a function of mixer RPM.*

RPM	q^I (cm³/s)	q^{II} (cm³/s)	E_f
0	3.7	13.0	0.71
0	3.7	8.5	0.98
0	2.4	7.6	0.82
0	2.5	6.5	0.51
0	2.4	8.4	0.89
0	2.4	8.4	0.95
0	2.4	8.4	0.87
0	2.4	8.4	0.92
0	3.9	18.2	0.93
182	3.9	18.2	0.97
235	3.9	18.2	0.97
305	3.9	18.2	0.98
370	3.9	18.2	0.98
500	3.9	18.2	0.99
505	3.9	18.2	0.98
580	3.9	18.2	0.97
630	3.9	18.2	0.99

10.7. In an experiment carried out by one of our undergraduate students, a water stream and a chloroform stream[**] were fed continuously to a 1,000 cm³ cylindrical tank with a cross-sectional area of 81 cm². The exit was located at one-fourth of the total height. Ammonia in amounts up to 1.7 g/L was dissolved in the chloroform

---

[*] This is a very good instructional problem, but in the half century since these data were obtained it was established that benzene is a human carcinogen. Hence, it would certainly not be used for this separation.

[**] This experiment was carried out more than forty years ago, before chloroform was classified as a toxic air contaminant and a probable human carcinogen. It should no longer be used for purposes like this one.

feedstream and extracted by the water stream. The chloroform stream is denoted as Phase I and the water stream as Phase II. The equilibrium distribution coefficient for ammonia is $M = c_{Ae}{}^I/c_{Ae}{}^{II} = 0.044$ at room temperature over the range studied. It can be assumed that the densities of the streams are independent of the ammonia concentration, with $\rho^I = 1{,}470$ kg/m^3, $\rho^{II} = 1{,}000$ kg/m^3. The fractional approach to equilibrium, E_f, was measured in the water effluent as a function of the mixer RPM, as shown in Table 10.P4.

    a.   At 0 RPM the phase interface in the tank is nearly a horizontal plane. Estimate the overall mass coefficient, K_m. Compare with the value obtained in the batch salt experiment in Section 10.3.3. What assumptions have you made?

    b.   Correlate $K_m a$ with RPM for this system, and in that way obtain a correlation for the efficiency of the mixer. Can you use your estimate of K_m from part (a) to obtain a meaningful correlation for the surface area?

    c.   Suppose that you wish to add another species to the aqueous phase that reacts with ammonia and is insoluble in chloroform. Can any of the terms in Equation 10.39 be neglected? Does this situation correspond to either of the limiting cases?

11 Equilibrium Staged Processes

11.1 Introduction

The implementation of a great many processes depends ultimately on the ability to separate various species from one another. In this chapter we will build on the basic principles of interphase mass transfer developed in Chapter 10 to explore some of the ideas involved in the design of a separation process. The essential component of the analysis is the *equilibrium stage* defined in Section 10.4.1.[*] The design problem can be roughly broken down into the actual mechanical implementation of an equilibrium stage and the computation of the number of equilibrium stages needed to effect a desired degree of separation. We shall deal only with the latter; the former remains very much an art and beyond the scope of this introductory text. For simplicity we will restrict this chapter to separation by liquid-liquid extraction. The overall approach and the solution techniques have much greater generality and are applicable to nearly all separation processes, phase-change and nonphase-change alike.

Liquid-liquid extraction is a process in which a solute is transferred between two solvents. We will assume that the two solvents are absolutely immiscible in one another; this is a convenient approximation, although only sometimes realistic. It is traditional to do separation process calculations using mass fractions instead of concentrations. We shall follow this practice, and Section 11.2 is devoted to reformulating the description of an equilibrium stage in terms of mass fractions.

11.2 Equilibrium Stage

A mass transfer stage for extraction is shown in Figure 11.1. The two phases with dissolved solute A are fed to a well-stirred contactor, where the agitation is designed to create a large interfacial area and transfer of A across the interface occurs. In the separator the phases separate because of density differences. An equilibrium stage has a sufficiently large residence time in the contactor to ensure that solute A

---

[*] This is not a serious restriction. Everything in this chapter can be repeated with minor adjustments for nonequilibrium stages with known stage efficiencies.

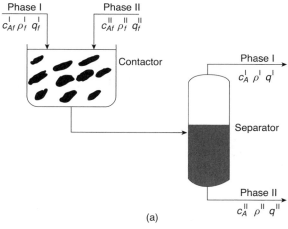

Figure 11.1. Continuous extraction stage: (a) nomenclature of Chapter 10; (b) traditional extraction nomenclature.

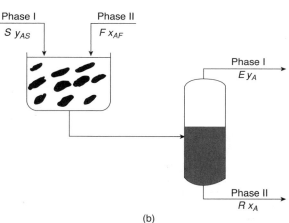

reaches its equilibrium distribution between the phases. Figure 11.1(a) is identical to Figure 10.4 and shows the nomenclature used in Chapter 10. Figure 11.1(b) shows the nomenclature traditionally used in extraction, as follows:

S: solvent (Phase I) mass flow rate $\quad S = \rho_f^I q_f^I,$
F: feed (Phase II) mass flow rate $\quad F = \rho_f^{II} q_f^{II},$
E: extract (Phase I) mass flow rate $\quad E = \rho_f^I q^I,$
R: raffinate (Phase II) mass flow rate $R = \rho_f^{II} q^{II}.$

The term *raffinate* was introduced in Chapter 5 in the context of membrane separation; y_A and x_A denote the mass fractions of solute A in Phases I and II, respectively. Since c_A denotes the mass of A per volume and ρ denotes the total mass per volume, the relations between concentrations and mass fractions must be as follows:

$$y_A = c_A^I/\rho^I, \quad y_{AS} = c_{Af}^I/\rho_f^I, \quad x_A = c_A^{II}/\rho^{II}, \quad x_{AF} = c_{Af}^{II}/\rho_f^{II}.$$

The equilibrium expression for solute concentrations in the two phases was written in Equation 10.8 as $c_A^I = Mc_A^{II}$. (We will not use the subscript e to denote equilibrium

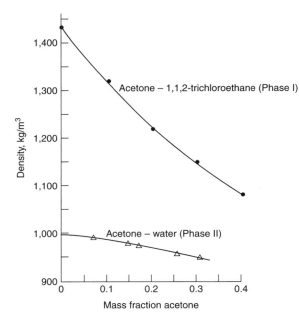

Figure 11.2. Density of acetone-1,1,2-trichloroethane and acetone-water solutions. Data of R. E. Treybal, L. E. Weber, and J. F. Daley, *Ind. Eng. Chem.*, **38**, 817–821 (1946).

since *all* concentrations in this chapter are equilibrium values.) In terms of mass fractions the equilibrium expression becomes

$$y_A = K x_A, \quad K = \frac{M \rho^{II}}{\rho^{I}}. \tag{11.1}$$

In general, K need not be a constant. Density data for solutions of acetone in water and in 1,1,2-trichloroethane are shown in Figure 11.2, and using these data the equilibrium data for this three-component system shown in Figure 10.2 are replotted in Figure 11.3 as y_A versus x_A. (Actually, the data were originally reported in mass fractions and recalculated for Figure 10.2.) Notice that there is less curvature in the equilibrium plot in Figure 11.3 than in Figure 10.2, and a constant value $K = 1.5$ provides an excellent fit to the data over a fairly wide range of mass fractions.

We saw in Section 10.4.1 that the assumption that the phases are in equilibrium removes the need to consider each phase as a separate control volume, and we can take the entire stage as the control volume. There are three species: the solute A, the feed solvent, and the extracting solvent. We will assume steady state, in which case the overall mass balance equation is

$$0 = F + S - E - R. \tag{11.2}$$

This is simply the sum of Equations 10.27I and II rewritten in the nomenclature of this chapter. Application of conservation of mass to species A gives

$$0 = F x_{AF} + S y_{AS} - E y_A - R x_A. \tag{11.3}$$

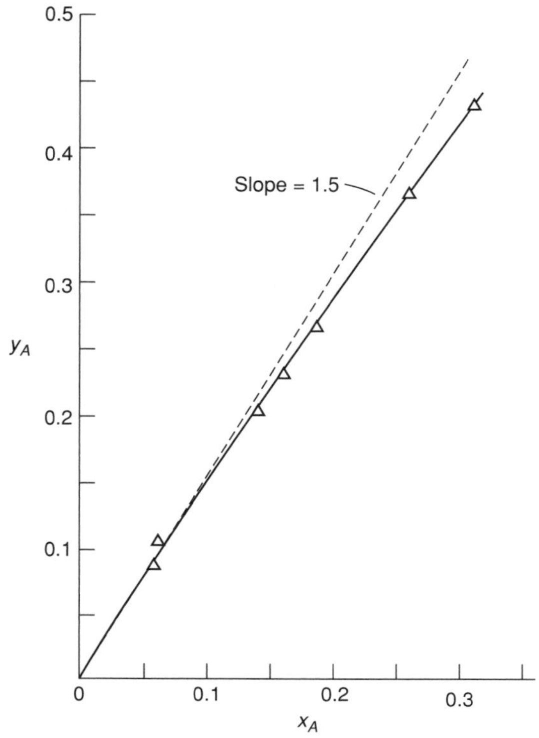

Figure 11.3. Equilibrium mass fraction of acetone in 1,1,2-trichloroethane (y_A) versus mass fraction of acetone in water (x_A). Data of R. E. Treybal, L. E. Weber, and J. F. Daley, *Ind. Eng. Chem.*, **38**, 817–821 (1946).

The assumption of immiscible solvents means that there are only two components for each phase, hence the mass fraction of solvent is simply one minus the mass fraction of A. Thus, for the feed solvent, conservation of mass requires

$$0 = F(1 - x_{AF}) - R(1 - x_A), \tag{11.4}$$

and for the extracting solvent

$$0 = S(1 - y_{AS}) - E(1 - y_A). \tag{11.5}$$

Only three of the four mass balance equations are independent, since Equation 11.2 is equal to the sum of Equations 11.3 to 11.5. Together with the equilibrium expression, Equation 11.1, these equations define the separation in an equilibrium stage.

11.2.1 Small Solute Transfer

The simplest situation to deal with is one in which the amount of solute transferred between phases is small. This approximation has wide applicability in practical situations since the error in using it is usually quite small, as we shall see in a subsequent section.

If the amount transferred is small, then $1 - x_{AF}$ and $1 - x_A$ are nearly equal, so Equation 11.4 reduces to $F \approx R$. Similarly, Equation 11.5 becomes $S \approx E$. These are simply statements that, despite the transfer of solute, the mass flow rates of Phases I and II are essentially unchanged. This approximation was made previously in deriving Equations 10.19I and II. Equation 11.3 then simplifies to

$$0 = F(x_{AF} - x_A) + S(y_{AS} - y_A). \tag{11.6}$$

Equation 11.6 is identical to Equation 10.29 in the changed nomenclature. By combining Equation 11.6 with the equilibrium expression, Equation 11.1, we can solve for x_A:

$$x_A = \frac{\Lambda x_{AF} + y_{AS}}{\Lambda + K}, \quad \Lambda = \frac{F}{S}. \tag{11.7}$$

Equation 11.7 is identical to Equation 10.32 in the nomenclature of this chapter.

EXAMPLE 11.1 An aqueous solution containing 200 kg/m³ of acetone is to be purified by continuous extraction with pure trichloroethane. How much acetone can be removed if the flow rates of both aqueous and organic streams are 0.01 m³/min (10 L/min)?

This example is identical to Example 10.1, except that here we have written everything in SI units. Since we have pure solvent, $y_{AS} = 0$. From Figure 11.3 we take $K = 1.5$. Calculation of x_{Af} requires trial and error, since the data in Figure 11.2 show ρ^{II} versus x_A, whereas we are given only $c_{Af}^{II} = \rho_f^{II} x_{AF}$. We will use the method of direct substitution to solve the nonlinear equation. As a first approximation we take $\rho_f^{II} \approx 1,000$ kg/m³, in which case $x_{AF} = c_{Af}^{II}/\rho_f^{II} \approx 200/1,000 \approx 0.2$. For $x_{AF} = 0.2$, $\rho_f^{II} \approx 970$ kg/m³, giving $x_{AF} = c_{Af}^{II}/\rho_f^{II} \approx 200/970 = 0.206$. Further calculation is unnecessary since ρ_f^{II} is essentially unchanged from the previous iteration. $\rho_f^{I} = 1,430$ for pure trichlorobenzene, so we can compute Λ as

$$\Lambda = \frac{F}{S} = \frac{\rho_f^{II} q_f^{II}}{\rho_f^{I} q_f^{I}} = \frac{970 \times 10^{-2}}{1,430 \times 10^{-2}} = 0.68.$$

Then

$$x_A = \frac{\Lambda x_{AF}}{\Lambda + K} = \frac{0.68 \times 0.206}{0.68 + 1.50} = 0.064.$$

From Figure 11.2, $\rho^{II} = 990$. Thus

$$c_A^{II} = \rho^{II} x_A = 990 \times 0.064 = 63 \text{kg/m}^3 = 63\text{g/L}.$$

In Example 10.1 we computed $c_A^{II} = 66.7$ g/L. This is reasonably good agreement, and the difference reflects inaccuracies in reading data from graphs, as well as roundoff in retaining only two significant figures in the calculation. Two significant figures is all that is reasonable here, since ρ^{II} cannot be determined from the graph with any greater accuracy.

11.2.2 Finite Solute Transfer

We need to use the full set of independent mass balance equations when the amount of solute transferred between phases is not small. Equations 11.4 and 11.5, together with the equilibrium Equation 11.1, can be written

$$R = F\left(\frac{1 - x_{AF}}{1 - x_A}\right), \quad E = S\left(\frac{1 - y_{AS}}{1 - Kx_A}\right).$$

Substitution into Equation 11.2 then gives

$$0 = F + S - S\left(\frac{1 - y_{AS}}{1 - Kx_A}\right) - F\left(\frac{1 - x_{AF}}{1 - x_A}\right).$$

With some slight rearrangements, using $\Lambda = F/S$, we obtain, finally,

$$\frac{(Kx_A - y_{AS})(1 - x_A)}{(x_{AF} - x_A)(1 - Kx_A)} = \Lambda \tag{11.8}$$

or, in standard quadratic form,

$$x_A^2 - \left[\frac{\Lambda - K\Lambda x_{AF} + K + y_{AS}}{K(\Lambda + 1)}\right] x_A + \frac{\Lambda x_{AF} + y_{AS}}{K(\Lambda + 1)} = 0. \tag{11.9}$$

Before solving Equation 11.9 for particular cases it is useful to examine the approximation involved in the small solute transfer assumption. We expand $(1 - x_A)/(1 - Kx_A)$ in a Taylor series about $x_A = 0$ to give

$$\frac{1 - x_A}{1 - Kx_A} = 1 + (K - 1)x_A + \cdots. \tag{11.10}$$

In the expansion in Equation 11.10 we have neglected the x_A^2 terms compared to x_A. For x_A of order 0.1, for instance, and K of order 2, the error is less than 1 percent. (Compare the discussion in Section 2.8.) Equation 11.8 can then be rearranged for x_A as

$$x_A = \frac{\Lambda x_{AF} + y_{AS}[1 + (K - 1)x_A + \cdots]}{\Lambda + K[1 + (K - 1)x_A + \cdots]}. \tag{11.11}$$

For K of order 2 and x_A of order 0.1, Equation 11.11 is equivalent to Equation 11.7 allowing for an uncertainty in y_{AS} and K of at most 10 percent. y_{AS} will generally be small or zero, whereas the experimental error in determining K may be comparable to the error introduced by the factor $1 + (K - 1)x_A + \cdots$. Thus, the small solute transfer equation can be expected to give reliable results even in many cases in which solute transfer does not really appear to be small.

EXAMPLE 11.2 Using the data in Example 11.1, compute x_A without the small solute transfer approximation.

We were given $y_{AS} = 0$, $K = 1.5$, and computed $x_{AF} = 0.206$, $\Lambda = 0.68$ from the available data. Equation 11.9 then becomes

$$x_A^2 - 0.95x_A + 0.056 = 0.$$

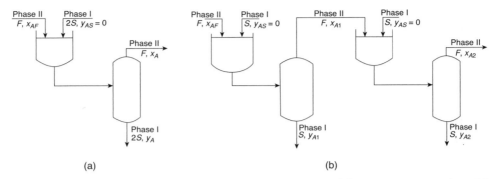

Figure 11.4. (a) One-stage extraction with solvent flow rate $2S$. (b) Two-stage extraction with solvent flow rate S to each stage. The total solvent used is the same in the two cases.

The positive root between zero and unity is $x_A = 0.063$. This is essentially the same as the mass fraction computed using the small solute transfer approximation, although 13 percent of the mass of the aqueous phase was transferred to the organic phase.

11.3 Two-Stage Extraction

The separation obtained in an equilibrium stage depends only on the separation factor K, the input mass fractions, and the relative mass flow rates. We showed in Examples 10.2 and 10.3 for a particular case that for a given system, fixed inputs, and a specified solvent flow rate, a better separation could be obtained by splitting the solvent between two successive stages than by using it all in one stage. This idea is familiar from the discussion of membrane separation in Chapter 5. We shall establish that result in general here for equilibrium stages and then extend it to see how the idea can be exploited with a substantial saving in solvent inventory. For algebraic simplicity it is assumed in everything that follows that solute transfer is small, phase mass flow rates are constant, and the extracting solvent is pure ($y_{AS} = 0$).

The two situations that we wish to compare are shown in Figures 11.4(a) and (b). The first configuration is the case that we considered in the preceding section, but we show the solvent stream as having a mass flow rate $2S$ to simplify later comparisons. In the second configuration we use the raffinate stream leaving the first stage as the feed to a second stage. Solvent mass flow is S in each stage so that the same total amount of solvent is used as in the single stage with which we are comparing. The subscript 1 is used for streams leaving the first stage and 2 for streams leaving the second. The configurations are shown schematically in Figures 11.5(a) and (b).

For the one-stage process the separation is determined by Equation 11.7. If we continue to define $\Lambda = F/S$ then, because we are using $2S$ for the solvent flow, for $y_{AS} = 0$ we obtain

$$x_A = \frac{\frac{1}{2}\Lambda x_{AF}}{\frac{1}{2}\Lambda + K},$$

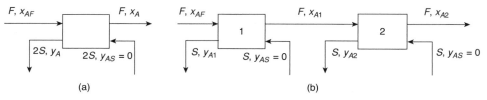

Figure 11.5. (a) Schematic diagram of one-stage extraction with solvent flow rate $2S$. (b) Schematic diagram of two-stage extraction with solvent flow rate S to each stage.

or, defining the one-stage separation ratio s_1,

$$s_1 = \frac{x_A}{x_{AF}} = \frac{\Lambda}{\Lambda + 2K}. \tag{11.12}$$

For the two-stage process, Figure 11.5(b), each stage is described by Equation 11.7. Thus,

$$x_{A1} = \frac{\Lambda x_{AF}}{\Lambda + K}.$$

The feed to the second stage has mass fraction x_{A1}, so Equation 11.7 gives

$$x_{A2} = \frac{\Lambda x_{A1}}{\Lambda + K}.$$

Combining these two equations we obtain the two-stage separation ratio, s_2:

$$s_2 = \frac{x_{A2}}{x_{AF}} = \frac{\Lambda^2}{[\Lambda + K]^2}. \tag{11.13}$$

It is straightforward to show that s_2 is always less than s_1, and hence that the two-stage process is more efficient, by taking the ratio:

$$\frac{s_2}{s_1} = \frac{\Lambda^2/[\Lambda + K]^2}{\Lambda/[\Lambda + 2K]} = \frac{\Lambda^2 + 2K\Lambda}{\Lambda^2 + 2K\Lambda + K^2} < 1. \tag{11.14}$$

If we use N consecutive stages, as shown in Figure 11.6, the separation ratio follows in a similar way as

$$s_N = \frac{\Lambda^N}{(\Lambda + K)^N}. \tag{11.15}$$

Compared to a single stage with solvent flow rate NS, the ratio of separation ratios is

$$\frac{s_N}{s_1} = \frac{\Lambda^N + NK\Lambda^{N-1}}{(\Lambda + K)^N} = \frac{\Lambda^N + NK\Lambda^{N-1}}{\Lambda^N + NK\Lambda^{N-1} + \frac{N(N-1)}{2!}K^2\Lambda^{N-2} + \cdots} < 1. \tag{11.16}$$

Multistage operation is an appealing alternative to using all of the solvent in one stage, since a better separation can be obtained with the same amount of solvent. There is still something wasteful about the crosscurrent process, however. In the first stage the mass fraction of solute is decreased to a fraction $\Lambda/(\Lambda + K)$ of that in the feed. In the next stage the further fractional reduction is the same. Thus, if

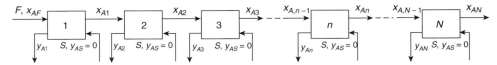

Figure 11.6. Schematic diagram of N-stage crosscurrent extraction with solvent flow rate S to each stage. Stage n denotes a typical stage.

50 percent of the solute is removed in the first stage, only 25 percent more is removed in the second, 12.5 percent in the third, and so on, although the same amount of solvent is required in each stage. The solvent leaving the latter stages will be quite dilute, and it may be necessary to handle large volumes of solvent containing only small amounts of solute to achieve a desired separation.

A nice compromise that combines the advantages of multistage operation with the use of a small amount of solvent readily suggests itself, as already discussed in Section 5.5 for membrane separation. Consider Figure 11.5(b). The extract stream leaving Stage 2 contains only a small amount of solute y_{A2}, because the greatest amount of solute was extracted in Stage 1. Thus, there would be little penalty if the solvent stream feeding Stage 1 were to contain mass fraction y_{A2} of solute instead of zero. This *countercurrent* operation is shown schematically in Figure 11.7. The raffinate stream from Stage 1 is the feedstream to Stage 2 and the extract stream from Stage 2 is the solvent stream for Stage 1. Equation 11.7, with appropriate nomenclature, applies to each stage. In the first stage the solvent stream contains mass fraction y_{A2}, so we have

$$x_{A1} = \frac{\Lambda x_{AF} + y_{A2}}{\Lambda + K}.$$

In the second stage the feed has mass fraction x_{A1}, and the solvent mass fraction y_{AS} is zero, so

$$x_{A2} = \frac{\Lambda x_{A1}}{\Lambda + K}.$$

These can be combined, together with the equilibrium relation $y_{A2} = Kx_2$, to give

$$s_{c2} = \frac{x_{A2}}{x_{AF}} = \frac{\Lambda^2}{[\Lambda + K]^2 - \Lambda K}. \tag{11.17}$$

s_{c2} denotes "separation ratio for two-stage countercurrent operation." Comparison of Equations 11.12 and 11.17 shows that as long as $K > \Lambda$ the separation in a two-stage countercurrent process is better than in a single stage that uses twice as much

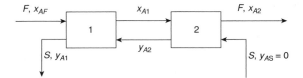

Figure 11.7. Schematic of two-stage countercurrent extraction.

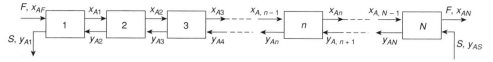

Figure 11.8. Schematic of N-stage countercurrent extraction.

solvent. Comparison with Equation 11.13 gives the relation between the separation ratios in countercurrent and cross-flow two-stage extraction:

$$\frac{s_{c2}}{s_2} = 1 + \frac{K\Lambda}{\Lambda^2 + K\Lambda + K^2}. \tag{11.18}$$

EXAMPLE 11.3   Using the data in Example 11.1, compute the separation for two-stage cross-flow and countercurrent operation.

We are given $\Lambda = 0.68$ and $K = 1.5$. Using Equations 11.13 and 11.17,

$$\text{cross-flow}: s_2 = \frac{x_A}{x_{AF}} = \frac{\Lambda^2}{(\Lambda + K)^2} = 0.097$$

$$\text{countercurrent}: s_{c2} = \frac{x_{A2}}{x_{AF}} = \frac{\Lambda^2}{(\Lambda + K)^2 - \Lambda K} = 0.123$$

In agreement with Equation 11.18, $s_{c2}/s_2 = 1.27$. The countercurrent operation uses 10 L/min of trichloroethane, whereas the cross-flow uses 20 L/min.

EXAMPLE 11.4   Using the data in Example 11.1, how much solvent will be needed to obtain a separation ratio of 0.097 in a two-stage countercurrent process?

According to Equation 11.17, we must solve for Λ from the equation

$$0.097 = \frac{\Lambda^2}{(\Lambda + K)^2 - \Lambda K}.$$

For $K = 1.5$ the positive root is $\Lambda = 0.58$. Since $\Lambda = F/S$ and $S = \rho_f^I q_f^I$, and we are given $F = 9.7$ kg/min and $\rho_f^I = 1{,}430$ kg/m^3, it then follows that $q_f^I = 0.0118$ m^3/min $= 11.8$ L/min. That is, 12 L/min in two-stage countercurrent operation gives a separation equal to that achieved using 20 L/min in two-stage cross-flow operation. Forty-two L/min of solvent would be required in a single stage.

11.4 Multistage Countercurrent Extraction

Design equations for a multistage countercurrent extraction process follow directly from the considerations in the preceding section. The process is shown schematically in Figure 11.8 with N stages. It is assumed that solute transfer is small, so the mass flow rates F and S are constant throughout all the stages. Consider Stage n, where n

is any stage between 2 and $N-1$. Conservation of mass applied to solute A leads to the equation

$$0 = Fx_{A,n-1} + Sy_{A,n+1} - Fx_{An} - Sy_{An}. \tag{11.19}$$

The equilibrium relation is still taken to be $y_A = Kx_A$, so

$$y_{An} = Kx_{An}, \quad y_{A,n+1} = Kx_{A,n+1}. \tag{11.20}$$

Combination of Equations 11.19 and 11.20 leads, with some rearrangement, to the analog of Equation 11.7:

$$x_{An} = \frac{\Lambda x_{A,n-1} + Kx_{A,n+1}}{\Lambda + K}. \tag{11.21}$$

It is convenient to introduce the ratio $\mu = \Lambda/K$, in which case we can write Equation 11.21 in the form

$$2 \le n \le N - 1: x_{A,n+1} - (1 + \mu)x_{An} + \mu x_{A,n-1} = 0. \tag{11.22}$$

For $n = 1$ and $n = N$ we have to take the input streams into account, and conservation of mass leads to

$$\text{Stage 1: } x_{A2} - (1 + \mu)\, x_{A1} + \mu x_{AF} = 0, \tag{11.23a}$$

$$\text{Stage } N: \frac{y_{AS}}{K} - (1 + \mu)\, x_{AN} + \mu x_{A,N-1} = 0. \tag{11.23b}$$

Equation 11.22, which relates the mass fraction x_A in successive stages $n-1$, n, and $n+1$, is called a linear finite difference equation. It can be manipulated together with Equations 11.23a,b in a manner identical to that used for linear differential equations, as shown in Appendix 5A, to obtain a solution for x_{An} explicitly in terms of N, μ, K, x_{AF}, and y_{AS}:

$$x_{An} = \frac{\left(x_{AF} - y_{AS}/K\right)\mu^n + y_{AS}/K - x_{AF}\mu^{N+1}}{1 - \mu^{N+1}}. \tag{11.24}$$

We shall not derive Equation 11.24, but we simply demonstrate that it is indeed a solution to Equation 11.22. By replacing n with $n-1$ and $n+1$, respectively, we can rewrite Equation 11.24 as

$$x_{A,n-1} = \frac{(x_{AF} - y_{AS}/K)\,\mu^{n-1} + y_{AS}/K - x_{AF}\mu^{N+1}}{1 - \mu^{N+1}}, \tag{11.25a}$$

$$x_{A,n+1} = \frac{(x_{AF} - y_{AS}/K)\,\mu^{n+1} + y_{AS}/K - x_{AF}\mu^{N+1}}{1 - \mu^{N+1}}. \tag{11.25b}$$

When Equations 11.24 and 11.25 are substituted into Equation 11.22 the terms do sum to zero, as required.

For algebraic simplicity we will consider from this point onward only the case of pure solvent, $y_{AS} = 0$. By setting $n = N$ in Equation 11.24 we then obtain an expression for the separation ratio in an N-stage countercurrent process:

$$s_{cN} = \frac{x_{AN}}{x_{AF}} = \frac{\mu^N - \mu^{N+1}}{1 - \mu^{N+1}}. \tag{11.26}$$

This equation can be rearranged to solve for N in terms of μ and s_{cN} to enable us to compute the number of equilibrium stages needed to achieve a given separation:

$$N = \frac{\log\left(\dfrac{s_{cN}}{1 - \mu + \mu s_{cN}}\right)}{\log \mu}. \tag{11.27}$$

Equation 11.27 is a version of what is often called the *Kremser Equation*.

EXAMPLE 11.5 Compute the separation in one-, two-, and three-stage countercurrent processes for $\Lambda = 0.68$ and $K = 1.5$ using Equation 11.26.

$$\mu = \frac{\Lambda}{K} = 0.45.$$

$$N = 1: s_{c1} = \frac{0.45 - 0.45^2}{1 - 0.45^2} = 0.31.$$

$$N = 2: s_{c2} = \frac{0.45^2 - 0.45^3}{1 - 0.45^3} = 0.12.$$

$$N = 3: s_{c3} = \frac{0.45^3 - 0.45^4}{1 - 0.45^4} = 0.05.$$

The result for $N = 2$ agrees, of course, with the calculation in Example 11.3.

EXAMPLE 11.6 Compute the number of equilibrium stages needed to remove 90 and 99 percent of the dissolved solute in the feedstream for $\mu = 0.45$.

The calculation is carried out using Equation 11.27. For 90 percent removal we have $s_{cN} = 0.10$ and

$$N = \frac{\log\left(\dfrac{0.10}{1 - 0.45 + 0.045}\right)}{\log(0.45)} = 2.2.$$

We cannot build 2.2 stages, of course, so we need three equilibrium stages to achieve 90 percent removal. From the calculation in Example 11.5 we see that we will, in fact, remove 95 percent of the solute with three stages. We might wish to decrease μ by a small amount by increasing solvent flow rate in order to carry out the desired separation in two stages. We saw in Example 11.4 that a 20 percent increase in S would achieve the removal of 90 percent of the solute in two stages.

For 99 percent removal we have $s_{cN} = 0.01$ and

$$N = \frac{\log\left(\dfrac{0.01}{1 - 0.45 + 0.005}\right)}{\log(0.45)} = 4.95 \cong 5.$$

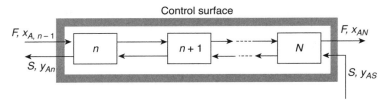

Figure 11.9. Control surface used for mass balances in developing the McCabe-Thiele method of graphical solution.

Twice as many stages are required to pass from 90 percent removal to 99 percent. This reflects the logarithmic dependence of the number of stages on the desired separation, which follows from Equation 11.27 and is discussed in Section 5.8 in the context of the marginal cost of separation.

11.5 Graphical Solution

The method leading to Equation 11.27 for computing the number of stages required to carry out a specified separation depends on the fact that all of the equations involved are linear in x_A and y_A. Nonlinearities can occur because the equilibrium relation is not linear or because changes in mass flow rates resulting from interphase transfer must be accounted for. Although the full set of nonlinear equations would normally be solved using a digital computer for precise results, a number of simple graphical procedures have been developed for rapid estimation, such as might be needed in preliminary design considerations. We shall discuss here the *McCabe-Thiele method*, a graphical method that is suitable when the assumption of constant mass flow rates is valid but the equilibrium is not linear. The McCabe-Thiele method is very useful for visualizing the computational process and for getting an intuitive sense of the details of the separation that is difficult to obtain from a table of numbers at the end of a computerized numerical calculation.

The development of the graphical method of solving the extractor equations centers around the choice of a control volume. Instead of doing the mass balance on the single stage n, the control volume is chosen to comprise stages n to N, as shown in Figure 11.9. Conservation of mass applied to component A in this control volume then becomes

$$0 = Fx_{A,n-1} + Sy_{AS} - Fx_{AN} - Sy_{An},$$ (11.28a)

or

$$y_{An} = \Lambda x_{A,n-1} - \Lambda x_{AN} + y_{AS}.$$ (11.28b)

In addition, we have the equilibrium relation,

$$y_A = f(x_A).$$ (11.29)

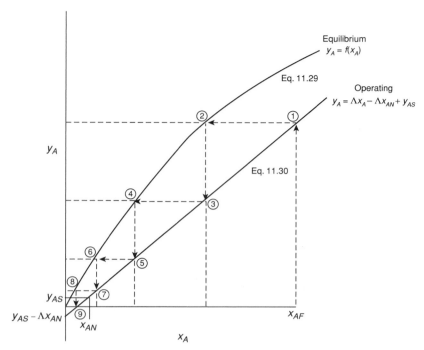

Figure 11.10. Calculation of the number of stages required for separation using the McCabe-Thiele method.

The construction is then carried out as shown on Figure 11.10. The equilibrium line, Equation 11.29, is drawn, as is the *operating line*,

$$y_A = \Lambda x_A - \Lambda x_{AN} + y_{AS}. \tag{11.30}$$

If we set $n = 1$, then $x_{A,n-1}$ is x_{AF}. According to Equation 11.28, y_{A1} is the value of the operating line when $x_A = x_{AF}$ (point 1). y_{A1} is in equilibrium with x_{A1}, so the equilibrium line, Equation 11.29, must give x_{A1} for $y = y_{A1}$ (point 2). Given x_{A1} we find y_{A2} from the operating line (point 3), then x_{A2} from the equilibrium line (point 4), and so on. We ultimately reach x_{AN} in this manner. The number of steps on the diagram corresponds to the number of stages required. Four stages would be needed in Figure 11.10, since x_{AN} lies between the third and fourth step.

EXAMPLE 11.7 For the equilibrium data shown in Figure 11.3, how many equilibrium stages are required to reduce the acetone mass fraction in the aqueous phase from 0.30 to 0.03 using $\Lambda = 0.68$ and pure solvent ($y_{AS} = 0$)?

The construction is shown in Figure 11.11. Between two and three (hence three) stages are needed. The equilibrium line is nearly linear in this range, so Equation 11.27 should provide a close approximation to the solution. In Example 11.6 a separation ratio of 0.10 was found to require 2.2 (hence three) stages for this system using the constant value $K = 1.5$.

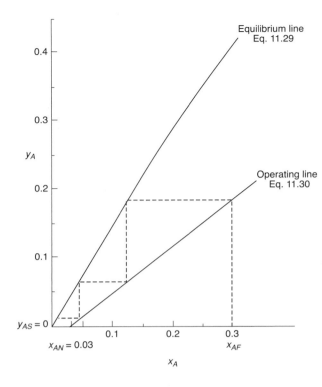

Figure 11.11. Calculation of the number of stages required to reduce x_A from 0.3 to 0.03 using $\Lambda = 0.68$ and the equilibrium data in Figure 11.3.

EXAMPLE 11.8 What is the minimum flow rate of pure solvent that can be used to reduce the acetone mass fraction of an aqueous phase from 0.30 to 0.03?

The solvent flow rate is contained in the ratio $\Lambda = F/S$, which is the slope of the operating line in the McCabe-Thiele method. The smaller S is, the larger is the slope. Figure 11.12 shows a sequence of operating lines with increasing slope. As Λ increases, the number of stages increases. The number of stages goes to infinity as the operating line intersects the equilibrium line at x_{AF}, and for Λ greater than this value the construction cannot be carried out. Thus, the minimum solvent flow corresponds to the slope Λ such that the two lines intersect at x_{AF}. For the given data and three-component system this is

$$\Lambda = \frac{F}{S_{min}} = 1.52, \quad S_{min} = 0.66F.$$

The desired separation thus requires that S be greater than $0.66F$. Beyond that point an economic balance between capital investment for the larger number of stages and the cost of handling larger volumes of solvent in the smaller number of stages must be carried out to arrive at the final design.

If the equilibrium line is nearly straight, as in the acetone-water-trichloroethane system, the calculation in Example 11.7 can be carried out algebraically with ease.

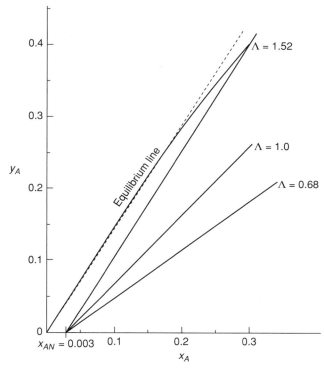

Figure 11.12. Calculation of the minimum solvent flow rate necessary to reduce x_A from 0.30 to 0.03.

From Equations 11.29 and 11.30 we see that the two lines intersect when $y_A = Kx_{AF} = \Lambda(x_{AF} - x_{AN}) + y_{AS}$, from which it follows that the maximum value of Λ, corresponding to the minimum value of S, is given by

$$\Lambda_{\max} = \frac{K}{1 - s_{cN}} - \frac{y_{AS}}{x_{AF} - x_{AN}}. \tag{11.31}$$

When $y_{AS} = 0$ the same result is obtained from Equation 11.27 by setting $1 - \mu + \mu s_{cN}$ to zero so that N goes to infinity, reflecting the fact that an infinite number of stages would be required to achieve the required separation at the minimum solvent flow rate.

EXAMPLE 11.9 Repeat the calculation of the minimum solvent flow rate in Example 11.8 with the assumption that the equilibrium line is linear with $K = 1.5$.

From Equation 11.31, with $s_{cN} = 0.1$ and $y_{AS} = 0$, we obtain $\Lambda_{\max} = 1.67$, hence $S_{\min} = 0.60F$. It is obvious from Figure 11.12 that taking the equilibrium slope as constant with an equilibrium curve that is concave will always lead to too low an estimate of the minimum solvent flow rate.

11.6 Concluding Remarks

The use of the equilibrium stage is not a serious restriction, since everything done in this chapter can be reformulated in nearly the same form for nonequilibrium situations by using the stage efficiency E_f introduced in Section 10.4.2. The essence of the analytical design question is expressed in Example 11.8: A certain minimum solvent flow rate is required for a given separation. Beyond the minimum there is a trade-off between costs associated with handling larger volumes and costs associated with adding additional stages. Given the capital and operating costs, the computation of an optimum follows in an identical manner to that for reactors in Chapter 7. The same comments apply to other separation processes, such as gas absorption, ion exchange, and distillation. Indeed, with appropriate changes of nomenclature to account for different physical processes, the equations and techniques developed in this chapter carry over to these and other separation processes almost without change. In particular, the logarithmic dependence of the number of stages on the desired degree of separation, as expressed in the Kremser Equation, is a common feature of all separations processes.

Selection of the solvent is a crucial issue in extraction, especially for high-value-added products like pharmaceuticals, which may be produced in small amounts. One interesting line of research undertaken by some chemical engineers in recent years has been to employ quantum mechanics to determine the interaction energy between complex molecules and potential solvents in order to evaluate solubility, utilizing computer codes that obtain approximate solutions to the time-independent Schrödinger equation. This approach will become more predictive when interaction energies obtained in this way are combined with statistical mechanical calculations that properly account for entropic contributions to the free energy of solvation. (These issues require a grounding in statistical thermodynamics and quantum mechanics, of course, both of which are areas of science utilized by many chemical engineers.)

The engineering art in separation begins with the choice of a process. We cannot deal with that crucial aspect of design here, for it is highly coupled with specific experience and with a thorough understanding of separations technologies. For difficult separations, in which a large number of stages is required, calculation uncertainty can be compensated for at minimal extra cost by adding extra stages, and this provides a useful safety factor if throughput must subsequently be increased because, say, of inaccurate market forecasts. The design of the configuration of a stage that will provide equilibrium, or at least high efficiency, is the significant engineering problem. This may involve the arrangement of baffling to induce adequate phase contact, the construction of efficient and inexpensive mechanical agitation for developing interfacial area, the design of phase separators that minimize the entrainment of one phase in the other, and so on. These considerations are beyond our analytical treatment here, and they still depend, to a large extent, on the designer's skill and appreciation of the particular system at hand.

Bibliographical Notes

The subject matter in this chapter is addressed in the unit operations books cited at the end of Chapter 10, as well as books specifically concerned with separations processes. Two recent examples of the latter are

> Henley, E. J., and J. D. Seader, *Separation Process Principles*, 2nd Ed., Wiley, New York, 2005.
>
> Wankat, P. K., *Separation Process Engineering*, 2nd Ed., Prentice-Hall, Englewood Cliffs, NJ, 2006.

The classic treatment of extraction is

> Treybal, R. E., *Liquid Extraction*, McGraw-Hill, New York, 1951; reprinted by Pierce Press, 2008.

PROBLEMS

The equilibrium data in Table 11.P1 for the distribution of acetic acid in water and isopropyl ether phases are used for Problems 11.1 through 11.4. x_A denotes the mass fraction of acid in the aqueous phase and y_A the mass fraction of acid in the organic phase.

Table 11.P1. *Equilibrium distribution of acetic acid between water (x) and isopropyl ether (y). Data of D. F. Othmer, R. E. White, and E. Trueger,* Industrial & Engineering Chem., *33, 1240–1248 (1941).*

x_A	0.007	0.014	0.029	0.064	0.133	0.255	0.367
y_A	0.002	0.004	0.008	0.019	0.048	0.114	0.216

11.1. It is necessary to process 100 kg/min of a water stream containing 15 percent acetic acid by weight. How much acid can be extracted in a single equilibrium stage with pure isopropyl ether using (a) 100 kg/min of solvent? (b) 50 kg/min? Assume negligible solute transfer between phases.

11.2. Repeat Problem 11.1 without the assumption of negligible solute transfer.

11.3. a. What is the minimum solvent flow rate that can be used in countercurrent operation to remove 75 percent of the acetic acid in the process stream described in Problem 11.1?

  b. How many countercurrent stages are required to remove 75 percent of the acid using a pure isopropyl ether solvent at a flow rate that is 150 percent of the minimum?

11.4. a. A separation is to be carried out in an N-stage cross-flow configuration using the same feed/solvent ratio Λ in each stage. The total feed/solvent ratio is therefore Λ/N. Compute the minimum solvent required for a given separation factor s_N. (That is, find the limit as $N \to \infty$.)

  b. Repeat the calculations in Problem 11.3a for this case and compare the minimum solvent requirements.

11.5. Trace amounts of phenol are to be extracted from an aqueous stream using pure xylene, for which $M = 1.4$ (*cf.* Problem 10.5). What is the minimum amount of xylene required to remove 90 percent of the phenol from the aqueous stream? Is this a feasible process?

11.6. The data in Table 11.P2 were obtained for the recovery of methyl ethyl ketone from water by solvent extraction. 500 kg/hr of an aqueous 18 percent by weight ketone solution is to be processed, with 85 kg/hr of ketone removed.

  a. Find the minimum flow rate of each solvent required to effect the separation. State any assumptions carefully.

  b. Find the number of equilibrium stages required for each solvent if (i) 1.2 times and (ii) 1.5 times the minimum solvent is to be used.

Table 11.P2. *Equilibrium data for methyl ethyl ketone in water and two organic solvents. Data of M. Newman, C. B. Hayworth, and R. E. Treybal,* Industrial & Engineering Chem., **41***, 2039–2043 (1949).*

Water-rich phase (wt %)			Solvent-rich phase (wt %)		
Ketone	Solvent	ρ (kg/m^3)	Ketone	Solvent	ρ(kg/m^3)
1,1,2-Trichloroethane as solvent					
18.15	0.11	970	75.00	19.92	890
12.78	0.16	980	58.62	38.65	970
9.23	0.23	990	44.38	54.14	1,006
6.00	0.30	990	31.20	67.80	1,114
2.83	0.37	990	16.90	82.58	1,126
1.02	0.41	1,000	5.58	94.42	1,136
Chlorobenzene as solvent					
18.10	0.05	970	75.52	20.60	860
13.10	0.08	980	58.58	39.28	900
9.90	0.12	980	43.68	55.15	950
7.65	0.16	990	29.65	69.95	990
5.52	0.21	990	17.40	82.15	1,030
3.64	0.28	990	8.58	91.18	1,060

12 Energy Balances

12.1 Introduction

Up to this point we have addressed physical situations for which mass is the only fundamental dependent variable, and by doing so we have been able to explore a wide range of chemical engineering applications. We have made some implicit assumptions about momentum and energy transport in doing so, however. When we assumed that vessels were perfectly mixed, with a consequent uniform concentration, we did not ask about the nature of mechanical agitation necessary to effect perfect mixing; to have done so would have required incorporation of momentum as a fundamental variable. Similarly, when we assumed that chemical reactors could be operated at a specified temperature and pressure, we did not consider the means by which temperature and pressure control could be effected, nor did we consider the possible impact of temperature transients or of the compressibility of a gas phase; to have done so would have required incorporation of energy as a fundamental variable.

Momentum transport is usually addressed in the chemical engineering curriculum in a course called Fluid Mechanics, or in the first part of a Transport Phenomena sequence. Energy is the subject of courses in Thermodynamics and Heat Transfer, where the latter may be incorporated in a subsequent part of a Transport Phenomena sequence, but an introduction to energy balances is often included in the first course for which this text is intended. We will therefore touch on energetics in order to illustrate the issues involved in incorporation of energy as a fundamental variable. We defer momentum transport in its totality to subsequent courses.

We all have an intuitive grasp of mass and how it is measured, hence it is usually obvious how the characterizing variables for mass should be selected. Selection of the characterizing variables for energy is not so readily done, because energy is not as familiar a concept as mass, and one's intuitive notion of energy may fail to incorporate all of the possible contributions to the total energy of a system. (Unfortunately, there is no device that serves as an energy meter.) Some forms of energy are familiar from basic courses in physics or mechanics, the most common being potential energy (*PE*) and kinetic energy (KE). The potential energy of a body of mass m is defined as the work necessary to raise the mass of the body to a given height L^* above a datum:

$$PE = \frac{mgL}{g_c}. \tag{12.1a}$$

g is the acceleration of gravity, which varies by about 0.4% on the earth, depending on latitude, but is usually taken to equal 9.81 m/s^2 in the SI system. g_c is the conversion factor required for dimensional consistency when both mass and force units are employed, as discussed in Appendix 2B.4, with an SI value of 1 kg m/Ns^2 and a value of 32.174 lb$_m$ ft/lb$_f$ s^2 in the Imperial Engineering system that is still in common use in the United States. g_c is often ignored when a metric (cgs or SI) system is employed because the magnitude is unity. Potential energy per unit mass, which we denote pe, is then

$$pe = gL/g_c. \tag{12.1b}$$

We have already introduced kinetic energy briefly in Section 2.9 in the context of the draining tank. Kinetic energy is defined as the work required to accelerate a constant mass m from rest to a velocity v, with a value

$$KE = \frac{1}{2}\frac{mv^2}{g_c}; \tag{12.2a}$$

the kinetic energy per unit mass, ke, is

$$ke = \frac{v^2}{2g_c}. \tag{12.2b}$$

The units of kinetic and potential energy per unit mass are J/kg and ft lb$_f$/lb$_m$ in the SI and Imperial Engineering systems, respectively.

12.2 Internal Energy

Potential and kinetic energy are familiar concepts that are defined in terms of an amount of work that must be done on a material to achieve a given height or a given velocity, respectively. Our experience with the physical world tells us that under certain conditions we can do work on a system and detect changes in the physical state under conditions where both the kinetic and potential energies remain constant. A common example is to compress a gas in a cylinder by pushing a piston (consider a bicycle pump with the nozzle end sealed); work has been done on the gas by moving the piston through some distance, and we detect a change in the pressure and perhaps also in the temperature, but there has been no change in either the potential or kinetic energy. It is clear, then, that there must be another form of energy change that is related in some way to the temperature and pressure of the system; we call such an energy the *internal energy* and denote it by the symbol U.

---

* We have previously used h to denote a height, and this is the best choice mnemonically. The symbol h is commonly used for another physical quantity in thermodynamics, however, as noted subsequently, and we need to change notation.

The physical motivation for defining the internal energy was provided by a series of experiments carried out by James Prescott Joule between 1840 and 1850 in which water was maintained in a well-insulated tank (i.e., there was no heat loss to the surroundings) and work was done on the system. Joule was a brewer by profession, as well as an amateur scientist, and he was exploring the possibility of replacing the brewery's steam engines with the newly invented electric motor. In his most famous experiment a paddle was immersed in the water and made to turn by a series of pulleys attached to a falling weight. The work done by the falling weight was easily calculated. Joule found that the temperature of the water increased as a result of the work done, with one pound of water increasing by one degree Fahrenheit for each 773 ft lb_f of work done on the system (i.e., 698 J/kg °C). The other experiments were as follows:

I. An electric current was generated by mechanical work and a coil carrying the current was immersed in water (838 ft lb_f/lb_m°F).
II. Mechanical work was performed to compress a gas in a cylinder that was immersed in water (795 ft lb_f/lb_m°F).
III. Mechanical work was done on two pieces of iron rubbed together beneath the water surface (775 ft lb_f/lb_m°F).

Within the accuracy that could be achieved at the time, the four distinct methods gave essentially the same value. (The currently accepted value is 778.) A process in which there is no heat loss to the surroundings is called an *adiabatic process*. Joule's experiments thus lead to the following statement, which has been verified many times since:

> *The change of a body inside an adiabatic enclosure from a given initial state to a given final state involves the same amount of work, whatever the means by which the process is carried out.*

If subscripts A and B refer to the initial and final states, respectively, then this statement is expressed symbolically as

$$W = -[U_B - U_A]. \tag{12.3}$$

W is the work, which is defined as positive if performed *by* the system.[*]

We know from experience that we can raise the temperature of a body without doing any work by bringing it in contact with another body at a higher temperature. We must therefore postulate a mode of energy transfer that is different from work, which we call *heat* and denote with the symbol Q; Q is taken by convention to be positive if heat is added to the system. If we perform a thought experiment and consider a batch process whereby a body both absorbs heat and performs work, then the change in internal energy can be expressed as follows:

$$U_B - U_A = Q - W. \tag{12.4}$$

---

[*] This is the convention that is used in most textbooks on thermodynamics. The opposite convention is sometimes used, however, wherein work is positive if performed *on* the system. The particular convention that is used is unimportant as long as consistency is maintained.

Equation 12.4 is a statement of what is commonly known as the First Law of Thermodynamics, which states that the change in the internal energy of a batch system is equal to the heat added to the system less the work done by the system. U is a *state function*, its value depending only on the initial and final states, while Q and W are modes of *transfer* of energy between one system and another.

Equation 12.3 provides us with an experimental means of measuring changes in internal energy, provided only that we have the means to measure the work done. Equation 12.4 provides a means of measuring the amount of heat transferred if we have been able to use the adiabatic experiment to measure $U_B - U_A$. Thus, we make the important observation that the definitions of both internal energy and heat are *constructive*, in that they are defined only in terms of measurable quantities. We note also that only relative values of internal energy can be determined. Thus, a description of a physical process can never depend on the absolute magnitude of the internal energy, but only on changes.

Finally, it is important to note that the internal energy U must also depend on volume and composition, and internal energy per unit mass, u, must depend on density (reciprocal of volume/unit mass) and the concentrations of the component species. We draw on our experience to use a very important fact without formal proof, namely that the temperature, density, and composition uniquely determine the pressure in a single phase. Thus, it is sufficient to know the temperature, density or volume, and composition to establish the internal energy. This observation plays an important role in applications of the principle of conservation of energy, and we shall return to it repeatedly. The formal proof is usually included in a thermodynamics course.

12.3 A General Energy Balance

The principle of conservation of energy states that energy is neither created nor destroyed.[*] The First Law of Thermodynamics is a statement of energy conservation in a batch system. The application of the principle of conservation of energy to flowing systems simply requires that there be no source or sink terms (i.e., energy creation or destruction) in the balance equation.

Our pedagogy for developing energy balances will be somewhat different from that used for mass balances, although the logical principles remain unchanged. We will develop an energy balance for a well-stirred flow system (one in which we allow mass to cross the control surface) and then simplify the general balance as we apply it to various problems. The system and control surface shown in Figure 12.1 are general enough to be of use in many problems in chemical engineering without becoming confusingly complex. The figure is schematic only; it represents some general system in which work can be done by or on the system (represented by the paddle) and in which heat can be added or removed (represented schematically by

---

[*] This is, of course, a nonrelativistic concept, and applies only under conditions where Newtonian mechanics are applicable. We will never be concerned with the relativistic equivalence between mass and energy, and $E = mc^2$ is not a part of the normal chemical engineering lexicon.

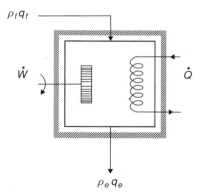

Figure 12.1. Schematic of a system for a general energy balance. Heat is added at a rate $\dot{Q}$ and work is done at a rate $\dot{W}$.

the coil). Material flows into the system at a mass flow rate $\rho_f q_f$ and out at a rate $\rho_e q_e$. The dot over the symbols W and Q denotes the rate. (The dot was Newton's symbol for a time rate of change, or a time derivative.) Note that we are changing our previous convention by using a subscript e to denote the stream leaving the tank (the effluent). Quantities in the tank remain unsubscripted. The reason will be apparent subsequently.

The energy balance is stated in words as follows:

The energy contained in the control volume at time $t + \Delta t$ must be equal to the energy contained in the control volume at time t, plus the total amount of energy that has appeared in the control volume in the interval Δt by all processes, less the total amount of energy that has disappeared from the control volume in the interval Δt by all processes.

We take the total energy to be the sum of internal energy, kinetic energy, and potential energy. Other forms of energy, such as surface energy and electromagnetic energy, are not considered, although they may be of importance in problems not considered in this text. Energy enters and leaves by convective flow; the energy is also increased by $Q = \dot{Q}\Delta t$, which is the total energy added as heat (schematically, through the coil), and the energy is decreased by $W = \dot{W}\Delta t$, which is the total energy removed as work (schematically, with the paddle). We then restate the energy balance in rate form as follows:

The rate of change of the total energy in the control volume equals the rate at which energy flows into the control volume, less the rate at which energy flows out of the control volume, plus the rate at which energy is added as heat flow across the control surface, less the rate at which energy is lost as work done by the system.

In equation form we then have

$$\frac{d[U + KE + PE]}{dt} = \frac{d[\rho V(u + ke + pe)]}{dt}$$
$$= \rho_f q_f (u_f + ke_f + pe_f) - \rho_e q_e (u_e + ke_e + pe_e) + \dot{Q} - \dot{W}. \quad (12.5)$$

It is usually convenient to separate the work term into two parts: Work must be done *on* the system to push material in the feedstream into the control volume, and work must be done *by* the system to push material out of the control volume in the effluent. All other work except this *flow work* (e.g., useful work for turning a turbine) is called *shaft work*. The rate of doing shaft work is designated $\dot{W}_s$. The flow work for the stream entering can be calculated as follows:

> *work* $= F\Delta l$, where F is force and Δl denotes the small distance an element of mass is pushed over the system boundary during a time interval Δt. $F = p_f A_f$, where p_f is the pressure at the entrance and A_f is the cross-sectional area. Thus, work $= p_f A_f \Delta l = p_f \Delta V$, where ΔV is the volume of the material pushed into the system during Δt. Then work/unit mass $= p_f \Delta V/\Delta$mass $= p_f/\rho_f$, since mass/volume is simply density. The rate of doing work is (work/mass) $\times$ (mass/time), or $(p_f/\rho_f)(\rho_f q_f)$. Finally, the term will enter the equation with a negative sign since work is being done on the system. Similarly, the rate of doing work *by* the system to push mass out is $(p_e/\rho_e)(\rho_e q_e)$.

Equation 12.5 then becomes

$$\frac{d[U + KE + PE]}{dt} = \frac{d[\rho V(u + ke + pe)]}{dt} = \rho_f q_f \left(u_f + \frac{p_f}{\rho_f} + ke_f + pe_f \right)$$
$$- \rho_e q_e \left(u_e + \frac{p_e}{\rho_e} + ke_e + pe_e \right) + \dot{Q} - \dot{W}_s . \tag{12.6}$$

Equation 12.6 can be used to obtain the flow rate relation for the draining tank considered in Section 2.7. Our focus on the use of energy balances is to understand thermal effects, however, and we do not wish to interrupt the logical flow in the body of the text. The draining tank is addressed in detail as an example in Appendix 12.A.

The combination $U + pV$ often appears in engineering applications and is given a special name, *enthalpy*, with the symbol H. Enthalpy per unit mass, h, then equals $u + p/\rho$, and Equation 12.6 can be written in an alternative form that is commonly employed:

$$\frac{d[U + KE + PE]}{dt} = \frac{d[\rho V(u + ke + pe)]}{dt} = \rho_f q_f (h_f + ke_f + pe_f)$$
$$- \rho_e q_e (h_e + ke_e + pe_e) + \dot{Q} - \dot{W}_s . \tag{12.7}$$

12.4 Heat Capacity

The internal energy and enthalpy must be related to characterizing variables that can be measured. Suppose we have a batch system in which there is no flow, so $KE = PE = 0$. Equation 12.6 then becomes

$$\frac{dU}{dt} = \dot{Q} - \dot{W} = \frac{dQ}{dt} - \frac{dW}{dt}, \tag{12.8}$$

which integrates directly to

$$\Delta U = \Delta Q - \Delta W, \tag{12.9}$$

where the symbol Δ, which denotes a difference, is intended here to indicate a small change. Equation 12.9 is, of course, equivalent to Equation 12.4, the First Law of Thermodynamics for a closed system.

We first consider a system in which the volume is kept constant while the heat is added. In that case no work can be done by the system and $\Delta W = 0$. We also assume that there is no chemical reaction, so the composition remains constant as the heat is added. We measure the small temperature change ΔT that results from the addition of the small amount of heat ΔQ at constant volume and *define* the heat capacity at constant volume, C_V, as

$$\Delta Q = C_V \Delta T = \Delta U \text{ at constant volume,} \tag{12.10}$$

or

$$C_V \equiv \lim_{\Delta T \to 0} \frac{\Delta U}{\Delta T} = \left. \frac{\partial U}{\partial T} \right)_{V, n_i}. \tag{12.11}$$

The partial derivative in Equation 12.11 is simply a symbol for the rate of change of internal energy with respect to temperature at constant volume and at constant composition (i.e., at a fixed number of moles of each component species). C_V can be determined experimentally by use of Equation 12.10, since both the heat input and the temperature change can be measured.

The heat capacity per se is not a useful quantity, because its value depends on the amount of material. The heat capacity per unit mass, c_V, which is a material property, is defined as

$$c_V = \left. \frac{\partial u}{\partial T} \right)_{\rho, c_i}, \tag{12.12a}$$

where the quantities held constant are density (reciprocal of volume/mass) and composition. It is sometimes convenient to work in terms of molar units. In that case we define an internal energy per unit mole as $\underset{\sim}{u}$ and define the heat capacity per unit mole as

$$\underset{\sim}{c}_V = \left. \frac{\partial \underset{\sim}{u}}{\partial T} \right)_{\rho, c_i}. \tag{12.12b}$$

Heat capacities per unit mass or per unit mole are often called *specific heats*.

We now consider a system that is maintained at constant pressure as heat is added. This can be done as shown schematically in Figure 12.2, where the cylinder is closed with a movable piston, so the volume can change if necessary to permit the pressure to remain constant. In that case W_s will not be zero, since the system will do work on the surroundings because the piston will move as the material expands with the addition of heat. We then define the heat capacity at constant pressure, C_p, as

$$Q = C_p \Delta T = \Delta U + \Delta W \text{ at constant pressure.} \tag{12.13}$$

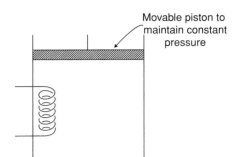

Figure 12.2 Schematic of a tank heated at constant pressure.

But the work ΔW equals the force acting on the piston multiplied by the distance over which the piston moves, which can be written it terms of the small volume change ΔV as $\Delta W = p\Delta V$, or, since the pressure is constant, $\Delta W = \Delta pV$. Hence, we can rewrite Equation 12.13 as

$$Q = C_p \Delta T = \Delta U + \Delta pV = \Delta(U + pV) = \Delta H \text{ at constant pressure}, \quad (12.14)$$

or

$$C_p \equiv \lim_{\Delta T \to 0} \frac{\Delta H}{\Delta T} = \frac{\partial H}{\partial T}\bigg)_{p,n_i}. \quad (12.15)$$

The heat capacities at constant pressure per unit mass, c_p, and per unit mole, $\underset{\sim}{c}_p$, are then defined as follows:

$$c_p = \frac{\partial h}{\partial T}\bigg|_{p,c_i}, \quad \underset{\sim}{c}_p = \frac{\partial \underset{\sim}{h}}{\partial T}\bigg)_{p,c_i}, \quad (12.16\text{a,b})$$

where $\underset{\sim}{h}$ is the enthalpy per unit mole.

A gas expands significantly with increasing temperature at constant pressure, so the heat capacities of gases at constant volume and at constant pressure will be different. Indeed, it is shown in physical chemistry texts that the molar heat capacities of a monotonic ideal gas (i.e., a gas for which $pV = nRT$, where n is the number of moles and $R = 8.3145$ J/mol K is the gas constant) at constant volume and at constant pressure are, respectively, $1.5R$ and $2.5R$. (For diatomic ideal gases $\underset{\sim}{c}_V = 2.5R$.) In contrast, liquids far from the critical point and solids expand very little, so the constant volume and constant pressure experiments are essentially the same. Hence, the heat capacities per unit mass at constant pressure and at constant volume are approximately equal for liquids and solids and it is usual not to distinguish between them.

Units of energy are briefly discussed in Appendix 2C. The calorie is defined as the heat required to raise the temperature of 1 gram of water by 1°C (without precise attention to temperature, but usually at 15°C), and the BTU is defined as the heat required to raise 1 lb_m of water 1°F (usually at 59°F). The specific heat of water is insensitive to temperature, so it is 1.0 over a wide range in both the cgs and Imperial Engineering systems. Most old data in the literature are tabulated in cal/g °C or BTU/lb °F, which are numerically equivalent. The SI unit is J/kg K, which

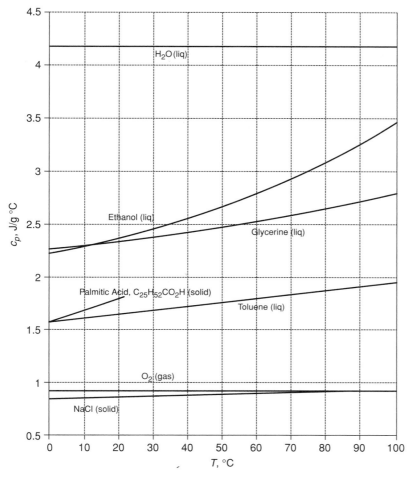

Figure 12.3. Temperature dependence of heat capacities at constant pressure of selected materials.

corresponds to 4,184 cal/g °C. Data are also commonly reported as J/g K; in these units the specific heat of water is 4.184. The temperature dependence of c_p of some typical materials is shown in Figure 12.3. Typical values for selected liquids and solids are shown in Table 12.1. The values for metals become closer when computed on a molar basis; the molecular weight of silver is nearly twice that of iron (the major component of steel), for example, while that of mercury is a bit less than twice that of silver. In fact, $c_{\sim V}$ for most metals is close to 25 J/mol K. (According to *the Law of Dulong and Petit*, $c_{\sim V} = 3R = 24.9$ J/mol K for a metal.) Typical values of both c_p and c_V for selected gases at 1 atm pressure are shown in Table 12.2. The values for gases also become closer when computed on a molar basis, with the diatomic gases in the table (H_2, O_2, N_2) all being close to the ideal gas value of $c_{\sim V} = 2.5R = 20.8$ J/mol K.

Table 12.1. *Heat capacities at constant pressure of selected solids and liquids.*

Material	State	$T\,(°C)$	c_p (J/kg K $\times 10^{-3}$) (J/g K)	$\underset{\sim V}{c}$ (J/g-mol K)
Silver	solid	20	0.23	24.8
Steel	solid	20	0.46	25.7
Sodium chloride	solid	0	0.84	49.1
Urea	solid	20	1.34	80.5
Water (ice)	solid	−10	2.22	40.0
Mercury	liquid	20	0.14	28.1
Sulfuric acid	liquid	20	1.42	139.3
Toluene	liquid	50	1.76	162.2
Ethylene glycol	liquid	15	2.39	148.3
Water	liquid	18	4.18	75.3

12.5 Temperature Equation for Liquid Systems of Constant Composition

We are now in a position to write the energy balance for a liquid system of constant composition in terms of the temperature, which is the characterizing variable of interest. (Pressure is also an important characterizing variable, but it plays little role for most liquid systems.) The development is straightforward but somewhat tedious. Our starting point is Equation 12.7. In most applications of interest to chemical engineers the kinetic and potential energy contributions are negligible relative to thermal effects, and we choose to ignore the *ke* and *pe* terms in the equation. (These terms are simply additive, so they can be added back if necessary.) Equation 2.7 then simplifies to

$$\frac{dU}{dt} = \frac{d\rho V u}{dt} = \rho_f q_f h_f - \rho_e q_e h_e + \dot{Q} - \dot{W}_s. \qquad (12.17)$$

The enthalpy depends on both temperature and pressure, but for a liquid system at moderate pressures we may ignore the pressure dependence. (See Appendix 12B.)

Table 12.2. *Heat capacities at constant pressure and constant volume of selected gases at 1 atm pressure.*

Gas	$T\,(°C)$	c_p (J/kg K $\times 10^{-3}$)	c_V	$\underset{\sim V}{c}$ (J/g-mol K)
Sulfur dioxide	15	0.63	0.50	32
Oxygen	15	0.92	0.67	21.4
Nitrogen	15	1.05	0.75	21
Ethylene	15	1.51	1.21	33.6
Water (steam)	100	2.01	1.51	27.2
Hydrogen	15	14.2	10.1	20.2

From Equation 12.16 we may therefore write, for a system at constant composition,

$$dh = c_p dT, \tag{12.18a}$$

or, for a change from temperature T_1 to T_2,

$$h_2 - h_1 = \int_{T_1}^{T_2} c_p dT = \text{(for constant } c_p) \quad c_p(T_2 - T_1). \tag{12.18b}$$

For simplicity in everything that follows we will assume that the heat capacity is independent of temperature, but this assumption is clearly unnecessary and is made just for convenience.

The temperature of the effluent is the same as the tank temperature, so we may drop the subscript from h_e. We need to evaluate all enthalpies at the same temperature, and the temperature of the tank is the most convenient choice, so we now use Equation 12.18b to express the enthalpy of the feedstream in terms of the tank and effluent enthalpy as

$$h_f = h + c_p(T_f - T). \tag{12.19}$$

Equation 12.17 then becomes

$$\frac{d\rho V \left(h - \frac{p}{\rho} \right)}{dt} = \rho_f q_f c_{pf} (T_f - T) + h (\rho_f q_f - \rho_e q_e) + \dot{Q} - \dot{W}. \tag{12.20}$$

In writing this equation we have put a subscript f on the heat capacity to emphasize that this is the heat capacity of the feedstream. This distinction is irrelevant now, but it will become necessary when we subsequently permit composition changes between feed and effluent. We have also replaced the internal energy u by $h - p/\rho$.

We now focus on the time derivative and write

$$\frac{d\rho V h}{dt} = \rho V \frac{dh}{dt} + h \frac{d\rho V}{dt} = \rho V \frac{dh}{dt} + h (\rho_f q_f - \rho_e q_e). \tag{12.21}$$

The second term on the right side of Equation 12.21 also appears on the right side of Equation 12.20. From Equation 12.18a we may write $dh/dt = c_p \, dT/dt$, so Equation 12.20 finally becomes

$$\rho V c_p \frac{dT}{dt} - \frac{dpV}{dt} = \rho_f q_f c_{pf} (T_f - T) + \dot{Q} - \dot{W}. \tag{12.22}$$

Finally, we show in Appendix 12B that the term dpV/dt can be neglected for liquid systems relative to the temperature derivative term, so we obtain the final form of the energy equation for a liquid system of constant composition as

$$\rho V c_p \frac{dT}{dt} = \rho_f q_f c_{pf} (T_f - T) + \dot{Q} - \dot{W}. \tag{12.23}$$

12.6 Concluding Remarks

The material in this chapter forms the basis for the treatment of all systems in which the energetics are important, which includes most systems that are of interest to chemical engineers, whatever their areas of application. Equation 12.23 is the prototypical form of the energy equation for liquid systems, and it is all that we need to analyze the heat transfer situations discussed in the next chapter. We will find that the equation is augmented when we have to address problems involving multiple species, whether reacting or simply mixing or being separated, because the internal energy and enthalpy will then depend on the composition, which may be changing during the course of the process. The pressure dependence of the enthalpy and the density dependence of the internal energy must be taken into account when dealing with gaseous systems, or with liquids that are near the critical point, both of which are beyond the scope that we wish to address in this introductory treatment. The approach developed here generalizes in a straightforward way, and it is the key to getting correct energy balances.

Bibliographical Notes

Joule's experiments were a landmark in developing modern concepts of energy. There is a late-nineteenth-century memoir by the great engineering scientist Osborne Reynolds (whom you will meet in fluid mechanics) that has been reprinted as

> Reynolds, O., *Biography of James Prescott Joule*, Wexford College Press, Palm Springs, CA, 2007.

For a more contemporary treatment, see

> Steffens, H. J., *James Prescott Joule and the Concept of Energy*, Dawson Publishing, Inglewood, CA, 1979.

The elementary principles of thermodynamics introduced here are usually treated in introductory courses in physics and chemistry, and they are elaborated on in courses in physical chemistry and in specialized courses in thermodynamics. Any text with the word *thermodynamics* in its title will cover these topics.

Physical property data are available from many sources, including handbooks such as *the Handbook of Chemistry and Physics* (CRC) and *Perry's Chemical Engineer's Handbook* (McGraw-Hill), which are updated with new editions periodically. The U.S. National Institute of Standards and Technology (NIST) provides an extensive compilation of physical properties, which is available electronically without cost. The URL for fluid properties was http://webbook.nist.gov/chemistry/fluid/ at the time of writing but, as with all Web-based information, it is important to check that this is the current site. Many academic institutions have access to the American Institute of Chemical Engineers' Design Institute for Physical Properties (DIPPR) electronic database of thermophysical properties. The classic *International Critical Tables of Numerical Data, Physics, Chemistry and Technology*, which was compiled between 1926 and 1930, can be found in most technical libraries and is also available in a facsimile electronic edition (Knovel, 2003).

PROBLEMS

12.1. Do a search to find an original research paper in which the heat capacity of some liquid was measured. Carefully read the paper and describe the experimental procedure in your own words. Show clearly by means of a sample calculation how the heat capacity was calculated from the experimental data.

12.2. What is the change in enthalpy of 1 kg of water when the temperature is changed from 25°C to 10°C?

12.3. The heat capacity of liquid toluene at atmospheric pressure is well represented in the range 280 K to 360 K by the empirical equation $c_p = -1.17 + 9.61 \times 10^{-3}T$, where T is in K and c_p is in J/g °C. What is the enthalpy change per gram in heating the liquid from 10°C to 60°C?

12.4. Repeat the calculation in Problem 10.3 with the assumption that the heat capacity of liquid toluene is constant, using the values at (a) 10°C; (b) 35°C; and (c) 60°C. How large is the error in each case? Give a quantitative explanation to the answer to part (b).

12.5. 10 kg of water is heated to raise the temperature by 7°C. How high would the liquid have to be lifted to change the total energy by the same amount? To what velocity must we accelerate the liquid to obtain the same change in total energy?

Appendix 12A: Draining Tank

We return here to the draining tank example of Section 2.7. The process is shown schematically in Figure 12A.1. The tank is assumed to be open to the atmosphere, so the air pressure above the liquid is atmospheric, and the effluent flows into an environment that is also at atmosphere pressure. The analysis depends on the choice of the control volume. As shown here, the control surface is drawn around the tank to the height of the liquid, and then across the tank at the liquid surface. With this choice of control volume there is no flow into the control volume, but the volume is changing with time. The alternative would be to draw a fixed control volume, in which case we would have to address the flow of air into the control volume as liquid flows out.

The temperature is fixed, so the liquid density is unchanged and can be denoted everywhere by ρ. There is no flow into this control volume, so $q_f = 0$. No heat is added and no work is done, hence $\dot{Q} = \dot{W} = 0$. The datum is taken at the tank exit, so $pe_e = 0$. The effluent jet is at atmospheric pressure (note that the pressure in the

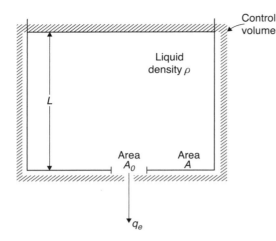

Figure 12A.1. Control volume for application of the energy balance to the draining tank.

jet is not the same as the pressure in the tank at the exit), so $p_e = 0$. Equation 12.6 thus simplifies to

$$\frac{d[\rho V(u + ke + pe)]}{dt} = (u + pe + ke)\frac{d(\rho V)}{dt} + \rho V\frac{d(u + ke + pe)}{dt} = -\rho q\,(u_e + ke_e).$$
(12A.1)

The equation of conservation of mass is $d(\rho V)/dt = -\rho q$. Also, $du/dt = 0$ since the temperature and density are constant, and $u = u_e$ since the temperature and density of the liquid in the tank and the effluent are the same. Hence, the internal energy terms on the two sides of the equation drop out. Every term is multiplied by the density, which may be factored out. $V = AL$. We thus obtain

$$L\frac{d(pe + ke)}{dt} = \frac{q}{A}\,(pe + ke - ke_e).$$
(12A.2)

The liquid in the tank is moving at a velocity q/A (volumetric flow rate/area), so $ke = \frac{1}{2g_c}\left(\frac{q}{A}\right)^2$. We recall from the study of mechanics in the introductory physics course that the potential energy of a finite mass is expressed in terms of the height of the center of mass, which in the case of the liquid in the tank is $L/2$, so $pe = gL/2g_c$. Finally, the velocity of the effluent is the volumetric flow rate/area of the exit jet. If the jet were to have the same area as the orifice this velocity would be q/A_o. In fact, it is known experimentally (and requires an application of the principle of conservation of momentum to derive theoretically) that the jet diameter below a sharp hole is only about 80 percent of the orifice diameter. Thus, the jet area is about 65 percent of the orifice area, and we write $ke_e = \frac{1}{2g_c}\left(\frac{q}{C_o A_o}\right)^2$, where C_o is known as the *orifice coefficient* and typically has a value between 0.6 and 0.65. A bit of manipulation of Equation 12.A2, which we leave as an exercise, then leads to the equation

$$2gL\left(1 + \frac{1}{g}\frac{d^2 L}{dt^2}\right) = \left(\frac{q}{C_o A_o}\right)^2\left[1 - C_o^2\left(\frac{A_o}{A}\right)^2\right].$$
(12A.3)

Equation 12.3 contains the second derivative of L. Integrating this equation to obtain L as a function of t would thus require two constants of integration. We know the initial height, but we do not know the initial velocity (although it probably does not differ significantly from zero). On the other hand, it is inconceivable that the acceleration of the liquid in the tank, which is what d^2L/dt^2 represents, is within orders of magnitude of the gravitational acceleration. Hence, we may safely assume that $\left| \frac{1}{g}\frac{d^2L}{dt^2} \right| \ll 1$ and neglect the acceleration term relative to unity. Similarly, $A_o \ll A$, so we may neglect the term $C_o^2 \left(\frac{A_o}{A} \right)^2$ relative to unity. We thus obtain the final result

$$q = C_o A_o \sqrt{2g L}. \tag{12A.4}$$

The square root dependence was found experimentally in Section 2.6.3, and a rather simplified version of this energy balance was used to obtain the correct scaling in Section 2.9. The functional form given here was obtained in Appendix 2D by dimensional analysis.

Appendix 12B: Pressure Dependence of Enthalpy

We justify here having neglected the pressure terms in the derivation of Equation 12.23 for liquids. It is shown in thermodynamics courses that the enthalpy change for small changes in temperature and pressure with a constant composition is

$$\Delta H = \rho V c_p \Delta T + \left[V - T\frac{\partial V}{\partial T} \right)_{p,n_i} \right] \Delta p. \tag{12B.1}$$

For most liquids far from the critical point, $V - T\partial V/\partial T \sim 10^{-3} VT$. Thus, $V - T\partial V/\partial T$ is less than V in magnitude, although perhaps of the same order. We therefore need to show that $V\Delta p$ is negligible compared to $\rho V c_p \Delta T$. Consider the ratio $V\Delta p/\rho V c_p \Delta T = \Delta p/\rho c_p \Delta T$. Typically, $\rho \sim 1{,}000$ kg/m^3 and $c_p \sim 2{,}000$ J/kg K (compare Table 12.1). If we take Δp to be of order 1 atm, or 10^5 N/m^2, then if ΔT is as small as 1 K we have $\Delta p/\rho c_p \Delta T \sim 10^5/(10^3 \times 2 \times 10^3 \times 1) \sim 5 \times 10^{-2}$. Thus, unless the pressure changes are very large, the pressure term is negligible in a liquid. The term multiplying Δp is identically zero for an ideal gas, but the term can be very important in gases at pressures and temperatures where there are deviations from ideal gas behavior.

The treatment of the pressure term in passing from Equation 12.22 to 12.23 is identical. If the volume is unchanged then the second term is identically zero at constant pressure. Otherwise we integrate $\rho V c_p \frac{dT}{dt} - \frac{dpV}{dt}$ over a short time interval to obtain $\rho V c_p \Delta T - \Delta p V$. The largest possible volume change is equal to V, in which case we seek to show that $\rho V c_p \Delta T$ is large compared to $V\Delta p$, which is precisely the comparison that we have already made.

13 Heat Exchange

13.1 Introduction

Most processes, regardless of scale, require heat exchange in order to control the temperature. This is true whether we are concerned with the operation of a microchip in a laptop computer, a 3,000-liter bioreactor, or the effect of the Gulf Stream on the weather of the North Atlantic. The basic concepts of elementary thermodynamics sketched out in the preceding chapter can be exploited to study heat exchange and temperature control for a wide range of applications of the types that we have considered previously in this text. We examine some simple but informative applications in this chapter. Combined mass, momentum, and heat transfer, together with chemical reaction, provide the foundation of the chemical engineering curriculum, so this treatment is just a beginning.

13.2 Rate of Heat Transfer

Consider the transfer of heat in the system shown in Figure 13.1. Two liquids at different temperatures are in adjacent well-stirred chambers. The tanks are completely insulated except for the surface separating the two liquids. There is no liquid flow in or out of either tank, and we assume that the shaft work to operate the agitators is unimportant ($\dot{W} \approx 0$). We take each tank individually as a control volume. In that case, Equation 12.23 for each tank simplifies, respectively, to

$$\rho_1 V_1 c_{p1} \frac{dT_1}{dt} = \dot{Q}_1, \tag{13.1a}$$

$$\rho_2 V_2 c_{p2} \frac{dT_2}{dt} = \dot{Q}_2. \tag{13.1b}$$

Because of the insulation, heat is transferred only between the two tanks. Hence, the heat entering one must equal the heat leaving the other, in which case

$$\dot{Q}_1 = -\dot{Q}_2. \tag{13.2}$$

Figure 13.1. Heat transfer between liquids in adjacent well-stirred chambers.

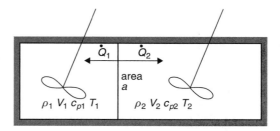

Equations 13.1a and b can then be combined to give

$$\rho_1 V_1 c_{p1} \frac{dT_1}{dt} = -\rho_2 V_2 c_{p2} \frac{dT_2}{dt}. \qquad (13.3)$$

The densities and heat capacities may be dependent on temperature. The analysis
that follows can be repeated for that situation as long as the temperature dependence
is known, but the manipulations are more complex and there is no fundamental gain
in understanding. We therefore assume that the physical properties of both liquids
are constant over the temperature range considered. In that case, Equation 13.3 is
simply an equality between time derivatives, so the integral of the left side must
equal the integral of the right. We take the temperatures of the tanks at time $t = 0$
to be T_{10} and T_{20}, respectively, and by integrating the derivatives from time $t = 0$ to
any later time we obtain

$$\rho_1 V_1 c_{p1}[T_1(t) - T_{10}] = -\rho_2 V_2 c_{p2}[T_2(t) - T_{20}]. \qquad (13.4)$$

We know that after a sufficiently long time the temperatures in the two tanks will
become equal, and we call this temperature T_∞. We can calculate T_∞ by setting both
T_1 and T_2 equal to T_∞ in Equation 13.4 and solving to obtain

$$T_\infty = \frac{\rho_1 V_1 c_{p1} T_{10} + \rho_2 V_2 c_{p2} T_{20}}{\rho_1 V_1 c_{p1} + \rho_2 V_2 c_{p2}}. \qquad (13.5)$$

Note the parallels in this development with phase concentrations in a two-phase
mass contactor and with the equilibrium in a single-phase batch reactor.

We will often be interested in the transient behavior, not just the long-time
steady-state temperature. To obtain the transient we need to solve Equation 13.1a
for $T_1(t)$. (If we know one temperature, the other is given by Equation 13.4.) In
that case, we need to know precisely what the rate of heat transfer $\dot{Q}_1$ is; that is, we
require a constitutive equation relating the rate of heat transfer to the temperatures
in the system. Now we find from experience that

  i) All other things being fixed, $\dot{Q}$ is proportional to the area, a, through which heat
     is transferred.
 ii) The cause of heat transfer is the temperature difference, and all other things
     being fixed, $\dot{Q}$ is roughly proportional to the temperature difference between
     the two bulk liquids.

Thus, the constitutive equation, written specifically for control volume 1, is of the form

$$\dot{Q}_1 = h_T a \left(T_2 - T_1 \right). \tag{13.6}$$

h_T is known as a *heat transfer coefficient*[*], with units of W/m^2 K (or BTU/hr ft^2 °F in the Imperial Engineering system). It is important to recognize that Equation 13.6 is a *definition* of the heat transfer coefficient, since the other quantities in the equation – area, temperatures, and heat flow – are known or can be measured (at least in principle). The heat transfer coefficient depends on the properties of the two liquids, the degree of agitation or other flow characteristics, and the materials of construction of the vessel wall. Note the close analogy to the development of membrane separation in Section 5.1 and to mass transfer in two-phase contactors in Section 10.3. As with the corresponding mass transfer phenomena, the overall heat transfer coefficient has distinct contributions from the fluid phases on each side of the barrier and from the wall itself, but for our purposes here it suffices to treat it as an experimentally measurable quantity that will be available through correlations and handbook tables. The detailed treatment of heat transfer coefficients, including the derivation of correlations for various geometries and flow conditions, is addressed in subsequent courses on heat transfer or transport processes. A linear relation like Equation 13.6 was first proposed by Isaac Newton, and is sometimes known as *Newton's law of cooling;* this is a misnomer, because Equation 13.6 is a definition, not a fundamental law of nature.

Equations 13.1a and 13.6 combine to give

$$\rho_1 V_1 c_{p1} \frac{dT_1}{dt} = h_T a (T_2 - T_1). \tag{13.7}$$

Note that the temperature in tank 1 rises $(dT_1/dt > 0)$ if $T_2 > T_1$ (heat flows from $2 \to 1$) and decreases $(dT_1/dt < 0)$ if $T_2 < T_1$ (heat flows from $1 \to 2$). With Equations 13.4 and 13.5, Equation 13.7 becomes, after some algebraic manipulation, an equation for T_1:

$$\frac{dT_1}{dt} = -h_T a \left[\frac{1}{\rho_1 V_1 c_{p1}} + \frac{1}{\rho_2 V_2 c_{p2}} \right] (T_1 - T_\infty). \tag{13.8}$$

T_∞ is defined by Equation 13.5. We have seen equations of this form many times; it can be formally separated to give

$$\frac{dT_1}{T_1 - T_\infty} = -h_T a \left[\frac{1}{\rho_1 V_1 c_{p1}} + \frac{1}{\rho_2 V_2 c_{p2}} \right] dt$$

and integrated to give the expected exponential approach to equilibrium:

$$\frac{T_1(t) - T_\infty}{T_{10} - T_\infty} = \exp \left(-h_T a \left[\frac{1}{\rho_1 V_1 c_{p1}} + \frac{1}{\rho_2 V_2 c_{p2}} \right] t \right). \tag{13.9}$$

---

[*]   The symbol h without the subscript T is more commonly used, but that usage would lead to confusion with the symbol for the enthalpy per unit mass. The symbol U is often used for the overall heat transfer coefficient, but U is the symbol for the internal energy and is not available.

It is informative to rewrite Equations 13.5 and 13.9 as

$$T_\infty = \frac{T_{20}\left[1 + \dfrac{\rho_1 V_1 c_{p1} T_{10}}{\rho_2 V_2 c_{p2} T_{20}}\right]}{\left[1 + \dfrac{\rho_1 V_1 c_{p1}}{\rho_2 V_2 c_{p2}}\right]} \rightarrow T_{20} \text{ for } V_1/V_2 \ll 1, \tag{13.10}$$

$$\frac{T_1(t) - T_\infty}{T_{10} - T_\infty} = \exp\left(-\frac{h_T a}{\rho_1 V_1 c_{p1}}\left[1 + \frac{\rho_1 V_1 c_{p1}}{\rho_2 V_2 c_{p2}}\right]t\right)$$

$$\rightarrow \exp\left(-\frac{h_T a}{\rho_1 V_1 c_{p1}}t\right) \text{ for } V_1/V_2 \ll 1. \tag{13.11}$$

That is, if one chamber is much larger than the other, its temperature is essentially unaffected by the heat transfer, and the transient is determined entirely by the properties of the smaller chamber. This result is not surprising in any way, but it is useful to see how it follows logically from the formulation.

There is a further estimate that is very useful. As we have seen before, a negative exponential is effectively zero when the argument reaches –3. Thus, if we define t_∞ as the time at which the system has effectively reached the steady-state temperature T_∞, it follows from Equation 13.11 that

$$t_\infty = \frac{3\rho_1 V_1 c_{p1}}{h_T a}\left[1 + \frac{\rho_1 V_1 c_{p1}}{\rho_2 V_2 c_{p2}}\right]^{-1} \rightarrow \frac{3\rho_1 V_1 c_{p1}}{h_T a} \text{ for } V_1/V_2 \ll 1. \tag{13.12}$$

Note that the time scales inversely with the surface-to-volume ratio a/V; that is, the larger the surface-to-volume ratio, the more rapid is the response.

EXAMPLE 13.1 To get a sense of magnitudes, suppose that we have a well-mixed cubic chamber with sides of 0.3 m (30 cm, or about 1 ft) that is immersed in a very large heating bath such that $V_1/V_2 \ll 1$, and there is heat transfer through all six faces of the cube. How long does it take for the temperatures to equilibrate?

We assume that the physical properties are approximately those of water, so $\rho \sim 1{,}000$ kg/m^3 and $c_p \sim 4{,}000$ J/kg K. The volume is 0.027 m^3 and the area is 0.54 m^2. Heat transfer coefficients vary greatly, depending on the liquids, the type of agitation, and the materials of construction. We take a value here of $h_T = 350$ W/m^2 K, which is reasonable based on what is reported in the literature, but which could vary substantially in either direction, depending on the system. We thus obtain an estimate of t_∞ from Equation 13.12 of about 1,800 seconds, or 30 minutes, for the temperature to equilibrate with the surrounding large bath. Reducing the length scale by a factor of six, to 5 cm, will reduce the time by the same factor, to about 300 seconds, or 5 minutes.

13.3 Heat Transfer to a Jacket

Heat transfer in a continuous-flow tank system is often achieved by passing the cooling or heating fluid through a jacket that has been placed around the vessel, as

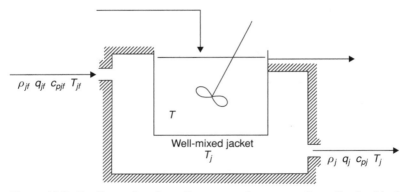

Figure 13.2. Cooling or heating a flow system by means of a well-mixed jacket.

shown in Figure 13.2. This is one common method for effecting temperature control in a flow reactor, for example. In this section we develop the design equations for sizing a jacket in order to achieve a specified rate of heat removal or addition. The fluid in the jacket is assumed to be well mixed and hence at a uniform temperature. We use the subscript j for the jacket fluid, and we leave the variables for the stirred tank unsubscripted.

We assume that there is no significant contribution from shaft work, so the energy balance for the tank is given by (*cf.* Equation 12.23)

$$\rho V c_p \frac{dT}{dt} = \rho_f q_f c_{pf} (T_f - T) + h_T a (T_j - T), \tag{13.13a}$$

where we have used Equation 13.6, written here as $\dot{Q} = h_T a (T_j - T)$, for the heat flow. The corresponding equation for the jacket is

$$\rho_j V_j c_{pj} \frac{dT_j}{dt} = \rho_{jf} q_{jf} c_{pjf} (T_{jf} - T_j) - h_T a (T_j - T), \tag{13.13b}$$

where $\dot{Q}_j = -\dot{Q}$. For simplicity, we will assume that the heat capacities and densities in the feedstreams are the same as those in the vessels, and that the flow rates in and out are the same, so we can drop the subscript f for all quantities except the temperatures.

At steady state, which is the operating condition for which we design, $dT/dt = dT_j/dt = 0$. We can then immediately solve Equation 13.13b for the required jacket temperature by setting the right-hand side to zero to obtain

$$T_j = \frac{T_{jf} + KT}{1 + K}, \tag{13.14}$$

where $K = h_T a / \rho_j q_j c_{pj}$. The required rate of heat transfer is then

$$\dot{Q} = h_T a (T_j - T) = \frac{h_T a}{1 + K} (T_{jf} - T). \tag{13.15}$$

It is useful at this point to refer back to the discussion of membrane separation in Chapter 5, where we noted that some useful limits on the feasible membrane area and the relative flow rates could be obtained rather simply from the design equation.

The same situation applies here, which is not surprising, given the obvious analogy between the formulations. For a given heat load $\dot{Q}$, the smallest surface area for heat transfer will occur when $|T - T_j|$ is as large as possible, which means that T_j should be as close as possible to the feed temperature T_{jf}. This will occur when $K \rightarrow 0$, corresponding to $q_j \rightarrow \infty$. Hence,

$$a_{\min} = \frac{\dot{Q}}{h_T \left| T_{jf} - T \right|}. \tag{13.16}$$

Similarly, the minimum coolant or heating fluid flow in the jacket will correspond to $a \rightarrow \infty$, in which case $K \rightarrow \infty$. With some algebraic manipulation of Equation 13.13b at steady state, combined with Equation 13.14, we then obtain in this limit

$$q_{j,\min} = \frac{\dot{Q}}{\rho_j c_{pj} \left| T_{jf} - T \right|}. \tag{13.17}$$

Equations 13.16 and 13.17 can be combined with the steady-state versions of Equations 13.13a and b to give

$$\frac{q_{j,\min}}{q_j} + \frac{a_{\min}}{a} = 1. \tag{13.18}$$

If, for example, we use twice the minimum area for heat transfer, we require a flow rate that is twice the minimum. If we use five times the minimum area, we require a flow rate that is only 25 percent greater than the minimum. The ultimate balance between heat transfer area and flow rate is an economic one, trading off capital outlay against operating costs for both the tank and the jacket. Of course, a is always confined within certain limits, since the volume of the tank, which dictates the surface area in contact with the jacket, will be determined by the requirements of the process taking place inside the tank.

EXAMPLE 13.2 Consider a heat transfer problem arising from an example of mixing water and an acid stream that is described in the next chapter. We wish to use an aqueous stream available at $T_{jf} = 20°C$ to remove heat from a tank at a rate of 6,250 W in order to maintain a tank temperature of 90°C. What are the specifications of the heat transfer system?

We can calculate the minimum heat transfer area required for the jacket from Equation 13.16: $a_{\min} = \dot{Q}/h_T \left| T_{jf} - T \right|$. If the heat transfer coefficient is 300 W/m^{2}K, which is a reasonable value for aqueous liquids in a steel tank, then we obtain a minimum area of 0.30 m^2. The minimum flow rate is computed from Equation 13.17: $q_{j,\min} = \dot{Q}/\rho_j c_{pj} \left| T_{jf} - T \right|$. Taking the density of the jacket fluid to be 1,000 kg/m^3 and the heat capacity to be 4.18 J/gK, we obtain a minimum flow rate of 2.1×10^{-5} m^3/s, or 1.26 kg of water per minute. If we use a heat transfer area of 1 m^2 then, according to Equation 13.18, the required coolant flow rate will be 3×10^{-5} m^3/s. This rate corresponds to 1.8 kg/min.

13.4 Heat Transfer to a Coil

Another design that is commonly used for heat transfer in a tank-type system is to flow coolant or heating fluid through a coiled pipe submerged in the tank, as shown in Figure 13.3. In this case, although the temperature in the well-mixed tank remains the same, the temperature of the fluid in the coil varies with position along the coil. Thus, the situation is analogous to the membrane system in Section 5.6, in which we must take the spatial variation of the characterizing variable into account by using a very small control volume.

The control volume is shown in Figure 13.4. The control volume is a length of pipe Δx long and D in diameter. Thus, x and $x + \Delta x$ correspond to the inlet and outlet of the control volume, respectively. The feed temperature is then $T_c(x)$, and the exit temperature is $T_c(x + \Delta x)$, where the subscript c is a mnemonic for "coil." The analysis uses the fact that the heat capacity and density change negligibly over the small distance Δx, which is certainly true as Δx goes to zero. We denote the rate of heat transfer to the coil over the small surface area $\pi D \Delta x$ as $\Delta \dot{Q}_c$. The fluid is assumed to be well mixed in this small volume, so we can write Equation 12.23 at steady state, with an obvious change of nomenclature, as

$$
\begin{aligned}
0 &= \rho_c q_c c_{pc} \left[T_c(x) - T_c(x + \Delta x) \right] + \Delta \dot{Q}_c \\
&= \rho_c q_c c_{pc} \left[T_c(x) - T_c(x + \Delta x) \right] + h_T \pi D \Delta x \left(T - T_c \right).
\end{aligned}
\tag{13.19}
$$

The coil temperature in the heat transfer term can be thought of as some average of the value in the control volume, or as the effluent value $T_c(x + \Delta x)$; it doesn't matter, since all temperatures will be the same when the length of the control volume is shrunk to zero.

We can now divide by Δx and rewrite Equation 13.19 as

$$
\frac{T_c(x + \Delta x) - T_c(x)}{\Delta x} = \frac{h_T \pi D}{\rho_c q_c c_{pc}} \left(T - T_c \right).
\tag{13.20}
$$

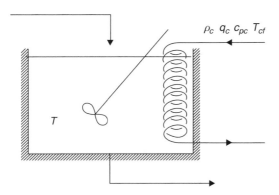

$\rho_c \ q_c \ c_{pc} \ T_{cf}$

Figure 13.3. Cooling or heating by means of a coil.

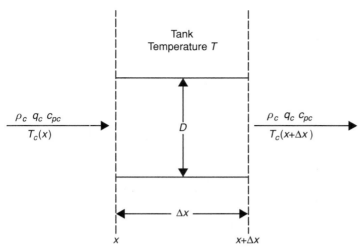

Figure 13.4. Control volume for a cooling or heating coil.

The left-hand side of the equation is a difference quotient that becomes the derivative dT_c/dx in the limit $\Delta x \to 0$, so we finally obtain an equation for the rate of change of the coil temperature with position:

$$\frac{dT_c}{dx} = \frac{h_T \pi D}{\rho_c q_c c_{pc}}(T - T_c).$$ (13.21)

An integration with respect to x will be required to find the temperature at each point, so we will need one constant of integration, namely that $T_c = T_{cf}$ at $x = 0$. We will assume that the density and heat capacity are insensitive to temperature and can be taken to be constants. Equation 13.21 is separable and can be integrated in the usual fashion to give

$$\frac{T_c - T}{T_{cf} - T} = \exp\left(-h_T \pi D x / \rho_c q_c c_{pc}\right).$$ (13.22)

Now, the rate at which heat is transferred between the tank and the coil over the small distance Δx is

$$\Delta \dot{Q}_c = h_T \pi D (T - T_c)\,\Delta x = h_T \pi D (T - T_{cf}) \exp\left(-h_T \pi D x / \rho_c q_c c_{pc}\right)\Delta x.$$ (13.23)

The total rate of heat transfer to the coil is the summation (integration) over all small distances:

$$\begin{aligned}
\dot{Q}_c &= \int_0^L h_T \pi D (T - T_{cf}) \exp\left(-h_T \pi D x / \rho_c q_c c_{pc}\right) dx \\
&= \rho_c q_c c_{pc}(T - T_{cf})\left[1 - \exp(-h_T \pi D L / \rho_c q_c c_{pc})\right] \\
&= \rho_c q_c c_{pc}(T - T_{cf})\left[1 - \exp(-h_T a / \rho_c q_c c_{pc})\right],
\end{aligned}$$ (13.24)

where $a = \pi D L$.

We might choose to define an overall heat transfer coefficient, say h_{TO}, in terms of the temperature difference $T - T_{cf}$; there is nothing stopping us from doing this, since the heat transfer coefficient is a defined quantity, and this temperature difference may be fixed by the design (the required operating temperature in the tank and the availability of cooling water at a fixed temperature, for example). We would then write $\dot{Q}_c = h_{TO}a\,(T - T_{cf})$, where $h_{TO} \equiv \frac{\rho_c q_c c_{pc}}{a}\,[1 - \exp(-h_T a/\rho_c q_c c_{pc})]$. It is quite common to use different driving forces in the definitions of heat transfer coefficients, and it is important when reading the literature to keep in mind the particular temperature difference that is used in the definition for any application.

It is worth pausing for a moment to reflect on the limiting behavior of the overall heat transfer coefficient h_{TO}. The dimensionless grouping $h_T a/\rho_c q_c c_{pc}$ reflects the ratio of heat transfer between the tank and the coil to energy flow through the coil. If this group is very small we make use of the fact that $1 - \exp(-y) \sim y$ for small y and find that h_{TO} becomes equal to h_T; that is, the rate of heat transfer is exactly the same as it would be in a quiescent system. If the group is large, on the other hand, the exponential term goes to zero and the rate of heat transfer is determined only by the flow rate in the coil; that is, the heat transfer across the barrier can be considered to be instantaneous, and the limiting factor is the rate at which the thermal energy can be carried away.

We can calculate the minimum area and minimum flow rate for the coil in the same way that we did in the preceding section for the jacketed tank. For the minimum area, corresponding to a flow rate approaching infinity, we again need to make use of the fact that $1 - \exp(-y) \sim y$ for small y. In that case, we recover Equation 13.16, but with T_{cf} in place of T_{jf}. Similarly, the exponential term vanishes as a becomes infinite, and we recover Equation 13.17 for the minimum flow rate with the same change in nomenclature. The relation between the minimum area and the minimum flow rate is now

$$\frac{a}{a_{\min}} = -\frac{q_c}{q_{c,\min}}\ln\left(1 - \frac{q_{c,\min}}{q_c}\right). \tag{13.25}$$

If we use a flow rate that is twice the minimum, for example, then we require a coil surface area that is 1.39 times the minimum $[\ln(0.5) = -0.693]$. This contrasts with the case of a jacket, where twice the minimum area is required for a flow rate that is twice the minimum. It can be established that a coil always requires a smaller area for the same heat load at a given flow rate and given jacket or coil inlet temperature. This result is intuitive when we consider that the driving force for heat transfer (the temperature difference) varies over the length of the coil, while it is constant everywhere for the well-mixed jacket.

EXAMPLE 13.3 Suppose we now use a coil to remove the heat in the system described in Example 13.2. What are the specifications of the heat transfer system?

   If we use a coil to remove the heat from the tank and we assume that the heat transfer coefficient is the same, then the minimum area and coolant flow rate are unchanged at 0.30 m^2 and 1.26 kg water/min, respectively. Now, however, if we use a heat transfer surface area of $1 m^2$ then, according to Equation 13.25, the required coolant flow rate will be 1.31 kg/min, which is just slightly more than the minimum and only 74 percent of the amount required for a jacket.

13.5 Double-Pipe Heat Exchanger

Much heat exchange takes place in *shell-and-tube* exchangers, the simplest of which is the double-pipe heat exchanger, shown schematically in Figure 13.5. Here, the hot fluid typically flows in the central pipe, whereas the cold fluid flows in an outer annulus, either co-currently or countercurrently. The development of the governing equations is the same for both co-current and countercurrent operation, but we choose the latter for specificity. The analysis is very straightforward but a bit tedious. Heat exchangers of this type are ubiquitous in a wide range of manufacturing operations in nearly all industries in which chemical engineers are employed, however, so it is important to get some early understanding of the way in which they work and, particularly, how the heat transfer calculations are done.

The control volumes, shown in Figure 13.6, are analogous to those used in Figure 5.5 for countercurrent membrane separation, and the approach directly parallels the treatment in Section 5.6. We use the subscripts t for the tube (the inner pipe) and s for the shell (the outer annulus in the double-pipe configuration), and we take separate control volumes of length Δx in the tube and shell, respectively. As in the preceding section, the rate of heat transfer across the interface with area $\pi D \Delta x$ from the tube to the shell is $h_T \pi D \Delta x (T_t - T_s)$, where D is the diameter of the tube. The tube fluid enters the control volume at x and exits at $x + \Delta x$, whereas the shell fluid enters at $x + \Delta x$ and exits at x. The energy balances at steady state for the tube and shell, respectively, are therefore

$$0 = \rho_t q_t c_{pt} \left[T_t(x) - T_t(x + \Delta x) \right] + \Delta \dot{Q}_t$$
$$= \rho_t q_t c_{pt} \left[T_t(x) - T_{ct}(x + \Delta x) \right] + h_T \pi D \Delta x \left(T_s - T_t \right), \qquad (13.26a)$$

$$0 = \rho_s q_s c_{ps} \left[T_s(x + \Delta x) - T_s(x) \right] + \Delta \dot{Q}_s$$
$$= \rho_s q_s c_{ps} \left[T_s(x + \Delta x) - T_s(x) \right] - h_T \pi D \Delta x (T_s - T_t). \qquad (13.26b)$$

For each of these equations we now follow the procedure employed in Section 5.6 for the countercurrent membrane separation, for which the formulation is identical in mathematical structure except for nomenclature, and in the preceding section for

Figure 13.5. Schematic of a double-pipe heat exchanger.

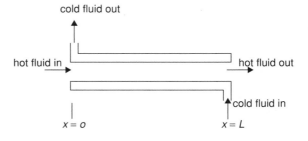

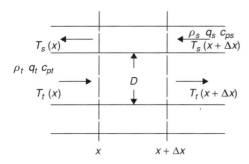

Figure 13.6 Control volumes for a double-pipe heat exchanger.

the coil equation; namely, we divide by Δx and take the limit as Δx goes to zero to obtain a pair of coupled equations:

$$\text{Tube}: \rho_t q_t c_{pt}\frac{dT_t}{dx} = h_T \pi D(T_s - T_t), \tag{13.27a}$$

$$\text{Shell}: \rho_s q_s c_{ps}\frac{dT_s}{dx} = h_T \pi D(T_s - T_t). \tag{13.27b}$$

The tube fluid enters with a temperature T_{to} at $x = 0$ and the shell fluid enters with a temperature T_{sL} at $x = L$. Again, following the procedure in Section 5.6, we assume constant physical properties for convenience, equate the left-hand sides of these two equations, and integrate the derivatives to obtain

$$T_t = T_{to} + \Lambda T_s - \Lambda T_{so}, \tag{13.28}$$

$$\beta\frac{dT_s}{dx} = T_s - T_t = (1 - \Lambda) T_s + \Lambda T_{so} - T_{to}. \tag{13.29}$$

The two parameters, which are introduced for notational convenience, are $\Lambda = \rho_s q_s c_{ps}/\rho_t q_t c_{pt}$ (dimensionless) and $\beta = \rho_s q_s c_{ps}/h_T \pi D$ (dimensions of length). Note that T_{so} is an unknown as we have formulated the problem (known feed temperatures, unknown outlet temperatures), and we have not yet used the fact that we know the feed temperature of the shell fluid, T_{sL}.

Equation 13.29 is of a form that we have solved many times, and we can readily establish that the solution in terms of the unknown quantity T_{so} is

$$T_s(x) = \frac{1}{1 - \Lambda}\left[(T_{so} - T_{to}) e^{(1-\Lambda)x/\beta} + T_{to} - \Lambda T_{so}\right]. \tag{13.30}$$

We can now establish the relation between the unknown T_{so} and the known temperature T_{sL} by setting $x = L$ in Equation 13.30 and doing some algebra to obtain

$$T_{so} = \frac{(1 - \Lambda)T_{sL} + \left[e^{(1-\Lambda)L/\beta} - 1\right] T_{to}}{e^{(1-\Lambda)L/\beta} - \Lambda}. \tag{13.31}$$

Equations 13.28, 13.30, and 13.31 contain all the information required to solve for the temperature profiles in the two parts of the heat exchanger. Typical temperature profiles are shown schematically in Figure 13.7. It is easy to show that the same equations hold for co-current flow, but now the known quantities may be presumed to be the two feed temperatures, T_{to} and T_{so}.

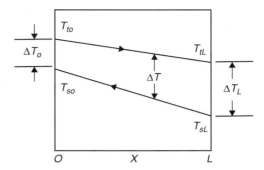

Figure 13.7. Typical temperature profiles in a countercurrent double-pipe heat exchanger.

For design purposes we are likely to need the amount of heat transferred between the two streams. As in Equation 13.24 for the coil in a tank, we write

$$\frac{\dot{Q}_s}{h_T \pi\, DL} = \frac{-\dot{Q}_t}{h_T \pi\, DL} = \frac{1}{L}\int_0^L [T_t(x) - T_s(x)]dx = \Lambda T_{so} + \frac{(T_{to} - T_{so})\,\beta}{(1 - \Lambda)\,L}\left(e^{(1-\Lambda)L/\beta} - 1\right)$$

$$= \frac{\Lambda}{e^{(1-\Lambda)L/\beta} - \Lambda}\left[(T_{to} - T_{sL})\left(\frac{\beta}{\Lambda L}\left(e^{(1-\Lambda)L/\beta} - 1\right) - 1\right) + e^{(1-\Lambda)L/\beta}\,T_{to} - T_{sL}\right].$$

$$(13.32)$$

Equation 13.32 provides the information that we want, but it is rather lacking in aesthetics and does not lend itself easily to the extraction of information. (It is an implicit equation for Λ, for example, which is the quantity needed to determine the flow rate ratio for a given heat load and a fixed area.) Equation 13.32 can be manipulated with considerable effort[*] into a form that is in common use:

$$\frac{\dot{Q}_s}{h_T \pi\, DL} = \frac{(T_{tL} - T_{sL}) - (T_{to} - T_{so})}{\ln\left[(T_{tL} - T_{sL})/(T_{to} - T_{so})\right]} \equiv \frac{\Delta T_L - \Delta T_o}{\ln\left(\Delta T_L/\Delta T_o\right)} \equiv \Delta T_{lm}. \quad (13.33)$$

ΔT_{lm} is known as the *log-mean temperature difference.*[**]

Equation 13.33 is a convenient form when the inflow and outflow temperatures of both streams are known and we wish to choose the area for a specified heat load, since the four temperatures and the heat load uniquely define the required heat transfer area for a given heat transfer coefficient. The most common design problem is one in which we are given the flow rate of a process stream, the inlet temperature, and the desired outlet temperature; for illustrative purposes we will take this stream

[*] The derivation of Equation 13.33 is much simpler if one seeks the equation directly. By subtracting Equation 13.37b from Equation 13.37a we obtain an equation for the difference $(T_t - T_s)$, which can be integrated to give $T_t(x) - T_s(x) = (T_{to} - T_{so})\exp(-(1 - \Lambda)x/\beta)$. An overall energy balance for which the control volume is the entire shell and tube between $x = 0$ and $x = L$ gives $T_{to} + \Lambda T_{sl} = T_{tL} + \Lambda T_{so}$. These two relations together immediately give Equation 13.33.

[**] It is important to note that the log-mean temperature difference is specific to the shell-and-tube heat exchanger and is defined in order to be able to write the overall heat transfer rate in the form of the total area multiplied by a heat transfer coefficient multiplied by a temperature difference. Students sometimes gain the mistaken impression in heat transfer courses that driving forces for heat transfer must always be of the log-mean form, which is, of course, untrue.

to be the tube flow. We are also typically given the inlet temperature of the cooling or heating medium, but we may or may not be given the flow rate.

The heat load is determined from an overall balance that takes the tube as the control volume, giving

$$\dot{Q}_s = \rho_t q_t c_{pt}(T_{to} - T_{tL}). \tag{13.34a}$$

When the heat load is known, the relation between the shell flow rate and the exit temperature T_{so} is determined from a balance that takes the shell as the control volume:

$$\dot{Q}_s = \rho_s q_s c_{ps}(T_{so} - T_{sL}). \tag{13.34b}$$

Equations 13.33 and 13.34a,b provide all of the information required to complete the design, provided that either the shell flow rate or the shell outlet temperature is specified. If neither is given then the problem is underdetermined and the design must include further criteria. Liquid pressure drop, which determines pumping costs, is often a factor that must be incorporated.

In practice, multiple parallel tubes (a "tube bundle") may be used in a single shell. It is straightforward to show that Equations 13.32 and 13.33 remain valid as long as D is replaced by ND, where N is the number of tubes in the bundle.

EXAMPLE 13.4 We must reduce the temperature of a small process steam flowing at 2 kg/min from 90°C to 30°C. The stream has the physical properties of water. Cooling water is available at 20°C and may be discharged at a maximum temperature of 50°C. What are the specifications of a double-pipe heat exchanger if $h_T = 400$ W/m²K?

The heat load is obtained from Equation 13.34a, with $c_{pt} = 4,180$ J/kg and $\rho_t q_t = 2$ kg/m $= 1/30$ kg/s, to be 8,360 W. If we assume that the cooling water is discharged at the maximum allowable temperature, 50°C, then it follows from Equation 13.34b that $\rho_s q_s = 1/15$ kg/s $= 4$ kg/min. $\Delta T_L = 30 - 20 = 10°$C and $\Delta T_o = 90 - 50 = 40°$C, hence $\Delta T_{\mathrm{lm}} = (10 - 40)/\ln(10/40) = 21.65°$C. From Equation 13.33, 8,360 W $= (400$ W/m²K$)\pi DL(21.65$ K$)$, and $\pi DL = 0.965$ m². If $D = 2.5$ cm, say, then $L = 12.3$ m. This is unrealistically long, but if ten parallel tubes of the same diameter were used then the length of the exchanger would be only 1.23 m.

13.6 Concluding Remarks

Temperature control is a critical element of almost everything that chemical engineers do. The modes of heat exchange introduced in this chapter give a sense of how things are done, and the design issues are fairly clear. Real heat exchangers, especially of the shell-and-tube type, tend to be more complex in design; there may be multiple tubes within a single shell, for example, or multiple passes of the tube within the shell, with the flow sometimes co-current and sometimes countercurrent. One or both fluids might be a gas or a boiling or condensing liquid. Texts on heat transfer and handbooks illustrate the range of possibilities and give design equations.

The general approach is the same for all designs, and the intellectually challenging issue is to understand the detailed contributions to the heat transfer coefficient from materials of construction, flow regimes, and fluid properties. This last aspect is usually a focus of the heat transfer portion of chemical engineering courses on transport processes.

Bibliographical Notes

Heat exchange is often covered in the chemical engineering curriculum as part of a sequence in transport processes (fluid mechanics and heat and mass transfer), or in a course on mass and energy balances, and there are many textbooks in common use with one or the other phrase in the title. Part of this chapter is based on material that first appeared in sections 11.8–11.10 of Russell and Denn, *Introduction to Chemical Engineering Analysis*. A recent text with some common material from Russell and Denn is

Russell, T. W. F., A. S. Robinson, and N. J. Wagner, *Mass and Heat Transfer: Analysis of Mass Contactors and Heat Exchangers*, Cambridge University Press, New York, 2008.

Mechanical engineering departments typically offer courses in heat transfer that may be included in the chemical engineering curriculum. There are many mechanical engineering textbooks on heat transfer, all of which include the material in this chapter. Chemical and mechanical engineering handbooks are good sources of illustrations of more complex heat exchange configurations and design equations.

PROBLEMS

13.1. Liquids are contained in concentric cylinders of height L, where the radius of the inner cylinder is R and the radius of the outer cylinder is ΛR. If the temperatures T_1 (inner) and T_2 (outer) are initially different, how long will it take to reach equilibrium? You may assume that the relevant physical properties of the two liquids are the same.

13.2. A cylindrical preheat tank with a length-to-diameter ratio of 1 and a capacity of 100 m^3 is used in a semibatch operation. The tank is completely jacketed, and the temperature of the heating material in the jacket is maintained at 100°C by steam that condenses. The material to be heated has the properties of water, and you may assume that $h_T = 800$ W/m^2 K. The initial temperature of the liquid in the tank is 20°C. How does the tank temperature vary with time? How long will it take for the tank temperature to reach 80°C?

13.3. You are required to design a jacketed system to heat a continuous flow of 1 m^3/min of a liquid with the physical properties of water from 20°C to 80°C. You may assume that the heating medium is condensing steam, so that the jacket temperature is always at 100°C. You decide to use a cylindrical tank with a length-to-diameter ratio of 1, and you may assume that that $h_T = 800$ W/m^2 K. How large must the tank be?

13.4. Show that Equations 13.32 and 13.33 apply to an exchanger in which there are N parallel tubes of identical diameter if D is simply replaced by ND.

13.5. Derive Equations 13.34a and b.

13.6. Consider the process in Example 13.4. Can you reduce the required area substantially by changing the coolant flow rate?

13.7. A crude oil stream available at 209°C is to be used to cool a heavy gas oil stream from 319°C to 269°C in a shell-and-tube heat exchanger. The crude oil is on the shell side and the gas oil on the tube side. The flow rates of the streams are 105,700 kg/hr for the gas oil and 367,600 kg/hr for the crude oil. The physical properties vary somewhat with temperature, but you may take the heat capacities of the gas oil and crude oil to be 3.14 and 2.67 J/kg K, respectively. The heat transfer coefficient is 450 W/m^2 K. What heat transfer area will be required? If the tube diameter is to be 15 cm, what size must the exchanger be? (The process variables and all parameters are taken from R. Mukherjee, *Chem. Eng. Progress*, February 1998.)

14 Energy Balances for Multicomponent Systems

14.1 Introduction

Most systems of interest to chemical engineers are multicomponent, and a bit of reflection tells us that the internal energy of a multicomponent system must depend on the composition as well as on the temperature and pressure. We know, for example, that the temperature increases without adding any heat when sulfuric acid and water at the same temperature are mixed together. Writing energy balances for multicomponent systems is straightforward, but it is delicate and requires a bit of care. The engineering literature, including textbooks and basic handbooks, abounds with incorrect energy balances, often because of unwarranted shortcuts. I have published a brief catalog of incorrect energy balances elsewhere.[*] Perhaps the most unsettling example on that list is a computer program offered for sale by a leading corporation to model a chemical reactor for converting coal to CO and H_2. The most important consideration in operating such a reactor is getting the location and magnitude of the highest temperature (the *hot spot*) right, because too high a temperature or an incorrect location will affect the structural integrity of the reactor. The model predicted steady-state and transient profiles of solid- and gas-phase temperatures, coal conversion, and the concentrations of many gaseous species, but it contained an error that guaranteed that the hot spot would be computed incorrectly!

The purpose of this chapter is to provide an introduction to energy balances for multicomponent systems. This is a subject that is addressed in detail in courses on thermodynamics, and also to some extent in courses on reaction engineering, and the scope is too broad to do more than touch on the subject here. As we have done previously, we will restrict ourselves to liquid systems far from the critical point; this enables us to ignore the pressure dependence of the enthalpy (*cf.* Appendix 12.3), which would not be permissible for a gas or for a liquid near the critical point, where

---

[*] See the Bibliographical Notes. Some years ago I gave a plenary talk entitled "How to write an incorrect energy balance, or Why should you be different from everyone else?" at an international conference on heat and mass transfer. The next speaker harrumphed a bit and then started his talk with "In view of the previous talk, the equations we used are not *exactly* correct, but...." Determining when commercial software is computing what it is supposed to compute may be the most difficult challenge faced by many engineering practitioners.

compressibility is important. This assumption, which is admittedly restrictive from the perspective of many applications, permits us to focus on the consequences of the composition dependence of the thermodynamic quantities and gives results that are easily generalized with the study of gas and multiphase systems at a later date.

14.2 Partial Molar Enthalpy

This section is tedious, but there is no choice other than to work through it slowly and carefully, because the concept that is introduced – the *partial molar enthalpy* – is essential to the analysis of the energetics of all multicomponent systems, from an industrial reactor to a cell in the human body.

Consider the enthalpy in a control volume. The enthalpy depends on the temperature and pressure, but also on the number of moles $n_1, n_2, \ldots, n_S$ of each of the S component species; that is,

$$H = H(T, p, n_1, n_2, \ldots, n_S). \tag{14.1}$$

Now, suppose that the temperature, pressure, and composition are changing with time. We then use the chain rule of differential calculus to write the rate of change of the enthalpy in terms of the rates of change of each of its arguments:

$$\frac{dH}{dt} = \left.\frac{\partial H}{\partial T}\right)_{p,n_i} \frac{dT}{dt} + \left.\frac{\partial H}{\partial p}\right)_{T,n_i} \frac{dp}{dt} + \left.\frac{\partial H}{\partial n_1}\right)_{T,p,n_i;\, i\neq 1} \frac{dn_1}{dt}$$

$$+ \left.\frac{\partial H}{\partial n_2}\right)_{T,p,n_i;\, i\neq 2} \frac{dn_2}{dt} + \cdots + \left.\frac{\partial H}{\partial n_S}\right)_{T,p,n_i;\, i\neq S} \frac{dn_S}{dt}. \tag{14.2}$$

Equation 14.2 is simply a mathematical statement of the obvious: The rate of change of the entire quantity is equal to the sum of the rates of change with respect to each argument, multiplied by the rate of change of that argument when all other arguments are held fixed.

The rate of change of the enthalpy with respect to each component species is called the partial molar enthalpy, denoted by the symbol $\tilde{h}_i$:

$$\tilde{h}_i \equiv \left.\frac{\partial H}{\partial n_i}\right)_{T,p,n_j;\, j\neq i}. \tag{14.3}$$

Imagine a beaker filled with a known composition at a given temperature and pressure. Now add an infinitesimal amount of species i while holding the temperature, pressure, and numbers of moles of all other component species constant. The partial molar enthalpy $\tilde{h}_i$ is the ratio of the infinitesimal change in the total enthalpy to the infinitesimal change in the number of moles of species i. We then write Equation 14.2 as

$$\frac{dH}{dt} = \left.\frac{\partial H}{\partial T}\right)_{p,n_i} \frac{dT}{dt} + \left.\frac{\partial H}{\partial p}\right)_{T,n_i} \frac{dp}{dt} + \tilde{h}_1 \frac{dn_1}{dt} + \tilde{h}_2 \frac{dn_2}{dt} + \cdots + \tilde{h}_S \frac{dn_S}{dt}. \tag{14.4}$$

The partial molar enthalpy is a *specific* quantity; that is, it is evaluated on a per mole basis. Hence, whereas the total enthalpy depends on the absolute number

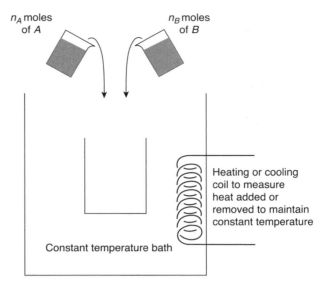

Figure 14.1. Schematic of a calorimeter experiment to measure the heat of solution.

of moles of each component species, the partial molar enthalpy depends on the molar *concentration* of each component species (and, of course, on temperature and pressure).

Now, let us do a thought experiment. Suppose that we start with an empty tank, and add liquid at a constant molar composition at a fixed temperature and pressure. The first two terms on the right-hand side of Equation 14.4 are therefore zero ($dT/dt = 0, dp/dt = 0$), and, because the composition is constant, the partial molar enthalpies $\tilde{h}_1, \tilde{h}_2, \ldots, \tilde{h}_S$ are constants. We can therefore write Equation 14.4 as

$$\text{Fixed } T, p, \text{ and } \{c_i\} : \frac{dH}{dt} = \frac{d}{dt}(\tilde{h}_1 n_1 + \tilde{h}_2 n_2 + \cdots + \tilde{h}_S n_S), \qquad (14.5)$$

or, integrating from the time at which we start filling the tank to the present,

$$H = \tilde{h}_1 n_1 + \tilde{h}_2 n_2 + \cdots + \tilde{h}_S n_S = \sum_{i=1}^{S} \tilde{h}_i n_i. \qquad (14.6)$$

The constant of integration is zero, since there is no enthalpy when the tank is empty. (Equation 14.6 is *not* an intuitively obvious result!) We can obtain the enthalpy per unit mass by dividing by ρV and using the fact that $c_i = n_i/V$:

$$h = \frac{1}{\rho}(\tilde{h}_1 c_1 + \tilde{h}_2 c_2 + \cdots + \tilde{h}_S c_S) = \frac{1}{\rho} \sum_{i=1}^{S} \tilde{h}_i c_i. \qquad (14.7)$$

14.3 Heat of Solution

The *heat of solution*, also called the *heat of mixing* (the *enthalpy change on mixing* is a better name), is a quantity that is often needed for calculations. The heat of solution is best defined through an experiment, which is illustrated in Figure 14.1.

We mix two pure steams together, and we carefully measure the heat that must be added or removed to keep the temperature constant. Such a device is known as a *calorimeter*. The control volume is the chamber in which the mixing takes place.

It is convenient to work in molar units. The mass balances for the two component species are then

$$M_{wA}\frac{dn_A}{dt} = \rho_A q_{Af}, \quad M_{wB}\frac{dn_B}{dt} = \rho_B q_{Bf}. \tag{14.8a,b}$$

M_{wi} denotes the molecular weight of species i. In writing the energy balance we need to generalize the development in Chapter 12 in an obvious way to include two inlet streams. There is no outlet stream, and we assume that there is negligible shaft work and that kinetic and potential energy terms can be neglected relative to the thermal terms. The generalization of Equation 12.7 to include two feedstreams with these assumptions is then

$$\frac{dU}{dt} = \rho_A q_{Af} h_A + \rho_B q_{Bf} h_B + \dot{Q}. \tag{14.9}$$

To reiterate, h_A and h_B are the enthalpies per unit mass of the pure species A and B, respectively, and the temperatures in the two feedstreams and in the tank are the same. The heat transfer term is required to maintain the tank temperature at the desired constant value.

We now make use of the fact that dU/dt is approximately equal to dH/dt for a liquid system far from the critical point. We also note that Equations 14.8a and b can be substituted for the ρq terms on the right side of Equation 14.9, and that $\underline{h}_i = M_{wi} h_i$ for i = A, B. Equation 14.9 then becomes, after a small amount of algebra,

$$\frac{dH}{dt} = \frac{d}{dt}(\tilde{h}_A n_A + \tilde{h}_B n_B) = \frac{d}{dt}(\underline{h}_A n_A + \underline{h}_B n_B + Q). \tag{14.10}$$

Q is the total amount of heat transferred, and $dQ/dt = \dot{Q}$. The integrals of the two sides from $t = 0$ until any future time must be equal, and the constant of integration must be zero, since initially there is nothing in the tank. We therefore write

$$Q = n_A(\tilde{h}_A - \underline{h}_A) + n_B(\tilde{h}_B - \underline{h}_B). \tag{14.11}$$

We will define A as the solute and B as the solvent. The heat added per unit mole of solute is known as the *integral heat of solution*, which we denote $\Delta \underline{H}_s{}^*$:

$$\Delta \underline{H}_s \equiv \frac{Q}{n_A} = (\tilde{h}_A - \underline{h}_A) + \frac{n_B}{n_A}(\tilde{h}_B - \underline{h}_B). \tag{14.12}$$

The heat of solution is thus measured directly in the calorimeter experiment. Typical data for a number of aqueous solutions at 25°C (298 K) are shown in Figure 14.2. If $\Delta \underline{H}_s > 0$ then, in accordance with the convention on the sign of Q, heat must be added to the system in order to keep the temperature constant; this is called *endothermic* mixing. If $\Delta \underline{H}_s < 0$, heat must be removed from the system to maintain the temperature; this is called *exothermic* mixing. A system for which the heat of

---

* For consistency with common usage, we have deviated here from our practice of using uppercase letters for total (extensive) quantities and lowercase letters for specific (intensive) quantities.

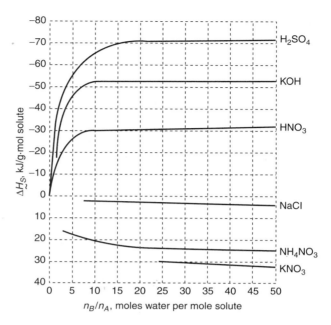

Figure 14.2. Heats of solution in water at 25°C. Selected values from Chemical Thermodynamic Properties, Nat. Bur. Standards Circ. #500 and supplements, 1952. The original data are in cgs units.

mixing is zero for all concentrations is called *ideal*; in this case the partial molar enthalpy of each species must be equal to the enthalpy per unit mole of the pure compound at the same temperature and pressure.

Consider the heat of solution data for the sulfuric acid–water system, for example, which we will use subsequently for some calculations. In our nomenclature, A is sulfuric acid (H_2SO_4) and B is water. The heat of solution of this system is strongly concentration dependent until a ratio of about 20 moles of water to 1 mole of sulfuric acid is reached. Beyond this concentration, $\Delta \underset{\sim}{H}_s$ becomes nearly constant at −73 kJ/mol acid, or −17.5 kcal/mol acid. This constant value is often called the *heat of solution at infinite dilution*. (A constant value clearly must be approached, because the partial molar enthalpy of water in the solution must approach the enthalpy per mole of pure water as the system become more and more dilute, and the second term on the right of Equation 14.12 will then go to zero.) The heat of solution at infinite dilution at 25°C is tabulated in standard collections of physical constants, and some representative data are shown in Table 14.1. In the absence of data at the needed concentrations, heat of solution data at infinite dilution can often be used to yield limiting values for preliminary design purposes.

EXAMPLE 14.1 To get a sense of magnitudes and the effect of concentration, suppose that we plan to mix 700 g of water with 100 g of sulfuric acid at 25°C, and we wish to maintain the temperature of the mixture at 25°C. How much heat must be removed?

The molecular weights of sulfuric acid (A) and water (B) are 98 and 18, respectively, so $n_A = 100/98 = 1.02$, $n_B = 700/18 = 38.9$, and $n_B/n_A = 38.1$. From Figure 14.2, $\Delta \underset{\sim}{H}_s = -73.4$ kJ/mole, and $Q = n_A \Delta \underset{\sim}{H}_s = 1.02 \times (-73.4) = -74.9$ kJ.

Table 14.1. *Heats of solution at infinite dilution in water at 25°C for selected compounds.*

Compound	ΔH_s at infinite dilution (kJ/g-mol)
Acetic acid	+9.6
Ammonium nitrate	+27.2
Cuprous sulfate	−48.5
Magnesium iodide	−209.8
o-nitrophenol	+26.3
p-nitrophenol	+18.8
Potassium hydroxide	−53.9
Potassium iodide	+21.7
Sodium chloride	+5.0
Sodium citrate	−22.2
Sucrose	+5.4

That is, 74,900 Joules (17,900 calories) must be removed to maintain the temperature at 25°C.

EXAMPLE 14.2 Now suppose that we plan to make 200 g of a 50-weight percent sulfuric acid–water solution, and then to mix that solution with 600 g of water to give us the same final concentration, and that we wish to keep the solution at 25°C. How much heat must be removed?

In the 50 percent solution $n_A = 100/98 = 1.02$, $n_B = 100/18 = 5.55$, and $n_A/n_B = 5.44$. At this molar ratio the integral heat of solution (to the accuracy at which it can be read from Figure 14.2) is –58 kJ/mol, and $Q = -58 \times 1.02 = -59$ kJ. Since the total heat that had to be removed to make the 700:100 g mixture was 74.9 kJ, the additional heat that must be removed on adding the additional 600 g of water to the 50:50 mixture is 15.9 kJ. Thus, it is clear that most of the heat removal is associated with the first addition of the water to the pure acid, and that there is a smaller marginal effect from further dilution. (On reflection, this conclusion is obvious from the shapes of the curves in Figure 14.2, but the concrete example is helpful in understanding the point.)

14.4 Heat Capacities of Mixtures

We need a few more relations before we can address problems of significance. The first task is to obtain an equation for the heat capacity of a mixture; for a binary mixture the heat capacity can be readily obtained in terms of the heat capacities of the pure materials and the enthalpy of mixing, as follows:

From Equation 12.14, $C_p = \frac{\partial H}{\partial T})_{p,n_i}$. By rearranging Equation 14.6 we can write

$$H = n_A \Delta H_S + n_A h_A + n_B h_B. \tag{14.13}$$

Thus,

$$C_p = \rho V c_p = n_A \frac{\partial \Delta \underset{\sim}{H_S}}{\partial T} + n_A \underset{\sim}{c_{pA}} + n_B \underset{\sim}{c_{pB}}, \tag{14.14a}$$

or

$$c_p = \frac{1}{\rho} \left[c_A \frac{\partial \Delta \underset{\sim}{H_S}}{\partial T} + c_A \underset{\sim}{c_{pA}} + c_B \underset{\sim}{c_{pB}} \right]$$

$$= \frac{1}{\rho} \left[c_A \frac{\partial \Delta \underset{\sim}{H_S}}{\partial T} + c_A M_{wA} c_{pA} + c_B M_{wB} c_{pB} \right]. \tag{14.14b}$$

For an ideal mixture, in which $\Delta \underset{\sim}{H_S}$ is zero, or in a nondeal mixture in which $\Delta \underset{\sim}{H_S}$ is a weak function of temperature, the heat capacity of the mixture is simply the weighted sum of the pure component heat capacities.

Heat of solution data as a function of temperature for a range of compositions are rarely available, so these equations are not especially useful. Equation 14.14b with $\partial \Delta \underset{\sim}{H_S} / \partial T$ set to zero (i.e., a weighted sum of pure component heat capacities) often provides an adequate estimate of the heat capacity of a mixture, although experimental data should be used whenever possible. When the heat of solution can be neglected, the heat capacity of a mixture of S species is

$$\text{Ideal solution: } c_p = \sum_{i=1}^{S} c_i \underset{\sim}{c_{pi}} = \sum_{i=1}^{S} c_i M_{wi} c_{pi}, \tag{14.15}$$

where $i = 1, 2, \ldots S$ refers to each of the component species in turn.

The results of this section can be used to calculate the temperature dependence of the heat of solution if the heat capacities are known, and this is the more useful result. Equation 14.14b can be written

$$c_A \frac{\partial \Delta \underset{\sim}{H_S}}{\partial T} = \rho c_p - c_A M_{wA} c_{pA} - c_B M_{wB} c_{pB}. \tag{14.16}$$

When the concentrations are constant, this equation can be integrated with respect to temperature from the temperature T^o at which data are reported to any temperature T to yield

$$c_A \Delta \underset{\sim}{H_S}(T) = c_A \Delta \underset{\sim}{H_S^o} + \int_{T^o}^{T} [\rho c_p(T') - c_A M_{wA} c_{pA}(T')$$

$$- c_B M_{wB} c_{pB}(T')] dT', \tag{14.17}$$

where T' is the dummy variable of integration and $\Delta \underset{\sim}{H_S^o}$ denotes $\Delta \underset{\sim}{H_S}(T^o)$. For constant heat capacities this simplifies to

$$c_A \Delta \underset{\sim}{H_S}(T) = c_A \Delta \underset{\sim}{H_S^o} + (\rho c_p - c_A M_{wA} c_{pA} - c_B M_{wB} c_{pB})(T - T^o). \tag{14.18}$$

14.5 Semibatch Mixing

Before addressing the general problem of continuous mixing it is helpful to return to the semibatch calorimeter shown schematically in Figure 14.1. The integral of Equation 14.10 can be written

$$H(T, t) = n_A(t)\underline{h}_A(T_A) + n_B(t)\underline{h}_B(T_B) + Q, \tag{14.19}$$

where $T(t)$ is the time-dependent temperature of the liquid in the tank and T_A and T_B are the constant temperatures of the pure inlet A and B streams, respectively. With Equation 14.13 we can then write

$$H(T, t) = n_A \tilde{h}_A(T, t) + n_B \tilde{h}_B(T, t) = n_A \underline{h}_A(T) + n_A \underline{c}_{pA}(T_A - T)$$
$$+ n_B \underline{h}_B(T) + n_B \underline{c}_{pB}(T_B - T) + Q, \tag{14.20}$$

where n_A and n_B are known functions of time, obtained directly from the rate at which each pure component is added. We have used the pure component heat capacities (assumed to be constants) to evaluate the pure component enthalpies at the tank temperature, T; this step is necessary because the various terms in the enthalpy of mixing must be evaluated at the same temperature. With the definition of the enthalpy of mixing, Equation 14.12, this equation can be rewritten as

$$T(t) = \frac{n_A \underline{c}_{pA} T_A + n_B \underline{c}_{pB} T_B + n_A[-\Delta \underline{H}_S(n_B/n_A, T)] + Q}{n_A \underline{c}_{pA} + n_B \underline{c}_{pB}}. \tag{14.21}$$

It is important to note that this equation is valid at all times, subject only to the assumption of constant pure component heat capacities and constant feed temperatures. The equation can be written in an equivalent form in terms of the enthalpy of mixing evaluated at the reference temperature by use of Equation 14.18 as

$$T(t) = T^o + \frac{n_A[-\Delta \underline{H}_S^o(n_B/n_A)] + n_A \underline{c}_{pA}(T_A - T^o) + n_B \underline{c}_{pB}(T_B - T^o) + Q}{\rho V c_p}.$$

$$\tag{14.22a}$$

ρV is the total mass in the tank at time t, and c_p is the composition-dependent heat capacity of the mixture. It is often convenient to write the equation entirely in terms of heat capacities per unit mass, as follows:

$$T(t) = T^o + \frac{x_A[-\Delta \underline{H}_S^o(n_B/n_A)]}{M_{wA} c_p} + x_A \frac{c_{pA}}{c_p}(T_A - T^o)$$

$$+ (1 - x_A)\frac{c_{pB}}{c_p}(T_B - T^o) + \frac{Q}{\rho V c_p}. \tag{14.22b}$$

x_A is the mass fraction of A at any time.

EXAMPLE 14.3 Suppose that we mix 50 kg each of sulfuric acid at –4°C (A) and water at 0°C (B) in an insulated vessel ($Q = 0$). What is the resulting temperature of the mixture?

After mixing we have $x_A = 0.5$, $M_{wA} = 98$, and $n_B/n_A = 98/18 = 5.4$. The enthalpy of mixing from Figure 4.1 at $T^o = 25°$ is about –58,500 kJ/g-mol acid.

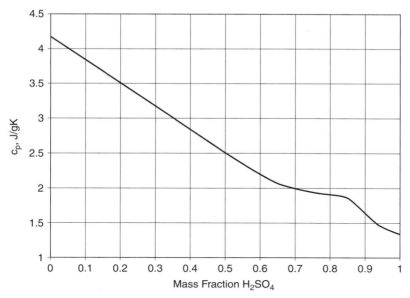

Figure 14.3. Heat capacity per unit mass of sulfuric acid-water mixtures at 20°C.

Heat capacity data for the sulfuric acid-water system at 20°C are shown in Figure 14.3, with values for the relevant streams as follows: water $= 4.2$ J/g K, 50 percent by weight acid $= 2.5$ J/gK, and 100 percent acid $= 1.4$ J/gK. We will assume for simplicity that the heat capacities are constant. Substitution of these numbers into Equation 14.22b, with $M_{wA} = 98$ and $Q = 0$, then gives a tank temperature of $T = 115$°C. This is below the boiling point of 50 percent sulfuric acid, which is about 123°C. Had we fed both streams at, say, 25°C, the computed temperature in the tank would have been 144°C, which is above the boiling point and outside the range of validity of the liquid analysis.

The large temperature rise on mixing is an important processing consideration. It is usually a poor idea to use constant values for the heat capacities over such large temperature ranges, but for the system at hand the error is not particularly serious.

EXAMPLE 14.4 We did not consider the manner in which the mixing was carried out in the preceding example, since we were interested only in the temperature when all of the material was in the tank. Now suppose we consider two different adiabatic mixing programs:

   I. The 50 kg of acid at –4°C is initially in the tank, and the water at 0°C is added at a constant rate of 5 kg/min for 10 minutes. Then $x_A = 50/(50 + 5t)$, $0 \leq t \leq 10$.
  II. The 50 kg of water at 0°C is initially in the tank, and the acid at –4°C is added at a constant rate of 5 kg/min for 10 minutes. Then $x_A = 5t/(50 + 5t)$, $0 \leq t \leq 10$.

We seek the temperature as a function of time for each program.

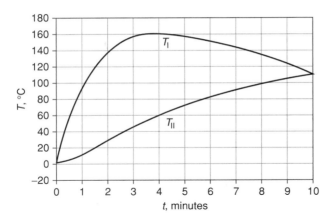

Figure 14.4. Temperature versus time. I: 50 kg of water at $0°C$ added at a rate of 5 kg/min to 50 kg of sulfuric acid initially at $-4°C$. II: 50 kg of sulfuric acid at $-4°C$ added at a rate of 5 kg/min to 50 kg of water initially at $0°C$.

The molar ratio n_B/n_A is computed at any time from the equation $n_B/n_A = M_{wA}(1 - x_A)/M_{wB}x_A$. $\Delta \underline{H}_S^o$ and c_p are obtained from Figures 14.2 and 14.3 at enough values of t between 0 and 10 to establish the two temperature versus time curves in Figure 14.4. Note that in Case II, acid added to water, the final temperature is approached gradually from below. In Case I, water added to acid, the temperature rises rapidly and reaches a maximum after 3 minutes that exceeds the final temperature by $47°C$, declining thereafter. The mixture does not boil in this case, since the boiling point of a mixture of mass fraction $x_A = 50/(50 + 5t)$ is always greater than the computed temperature $T_I(t)$ in this example. This calculation dramatically illustrates the reason for the chemistry laboratory safety rule of *add acid to water*. There are important aspects of the temperature transient that are best seen with a formulation that is developed in the next section, and we will return to this important example subsequently.

14.6 Continuous Mixing

We can now consider the general class of mixing problems shown in Figure 14.5. Streams 1 and 2, each mixtures of A and B at temperatures T_1 and T_2, respectively, are mixed in a well-stirred tank. We seek the equations describing the design and operation of such a system. The equations of conservation of mass are simply

$$\text{Overall mass: } \frac{d\rho V}{dt} = \rho_1 q_1 + \rho_2 q_2 - \rho q, \quad (14.23)$$

$$\text{Species A (moles): } \frac{dn_A}{dt} = \frac{dc_A V}{dt} = q_1 c_{A1} + q_2 c_{A2} - q c_A, \quad (14.24a)$$

$$\text{Species B (moles): } \frac{dn_B}{dt} = \frac{dc_B V}{dt} = q_1 c_{B1} + q_2 c_{B2} - q c_B. \quad (14.24b)$$

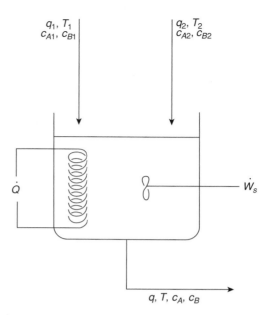

Figure 14.5. Continuous mixing of two streams, each containing A and B.

The energy balance, Equation 12.7, adjusted to account for two inlet streams and neglecting potential and kinetic energy terms and the difference between rates of change of U and H for a liquid system, is

$$\frac{dH(T)}{dt} = \rho_1 q_1 h_1(T_1) + \rho_2 q_2 h_2(T_2) - \rho q h(T) + \dot{Q} - \dot{W}_S, \qquad (14.25)$$

where we have been careful to note the temperature at which each enthalpy is evaluated. (Each enthalpy is also a function of the relevant composition, of course, but that is implied by the subscript or lack thereof.) We need to evaluate all enthalpies at the same temperature. There are two obvious choices: the temperature of the tank, and the reference temperature at which enthalpies of mixing are tabulated. We choose the former, in which case we can write

$$h_1(T_1) = h_1(T) + \rho_1 c_{p1}(T_1 - T),$$
$$h_2(T_2) = h_2(T) + \rho_2 c_{p2}(T_2 - T). \qquad (14.26a,b)$$

Here we have taken the heat capacities to be constants for convenience. In the case of temperature-dependent heat capacities we simply replace the second terms with integrals, as in Equation 12.18b.

Following Equation 14.4, we write Equation 14.25 as

$$\begin{aligned}
\frac{dH}{dt} &= \rho V c_p \frac{dT}{dt} + \tilde{h}_A \frac{dn_A}{dt} + \tilde{h}_B \frac{dn_B}{dt} \\
&= \rho V c_p \frac{dT}{dt} + \tilde{h}_A (q_1 c_{A1} + q_2 c_{A2} - q c_A) + \tilde{h}_B (q_1 c_{B1} + q_2 c_{B2} - q c_B) \\
&= \rho_1 q_1 h_1(T) + \rho_1 q_1 c_{p1}(T_1 - T) + \rho_2 q_2 h_2(T) \\
&\quad + \rho_2 q_2 c_{p2}(T_2 - T) - \rho q h(T) + \dot{Q} - \dot{W}_S.
\end{aligned} \qquad (14.27)$$

We now write (*cf.* Equation 14.7)

$$\rho_1 h_1 = c_{A1}\tilde{h}_{A1} + c_{B1}\tilde{h}_{B1}, \quad \rho_2 h_2 = c_{A2}\tilde{h}_{A2} + c_{B2}\tilde{h}_{B2},$$
$$\rho h = c_A\tilde{h}_A + c_B\tilde{h}_B. \tag{14.28a,b,c}$$

After a bit of algebra, in which the terms involving q cancel between the two sides of the equation, Equation 14.27 becomes

$$\rho V c_p \frac{dT}{dt} = \rho_1 q_1 c_{p1}(T_1 - T) + \rho_2 q_2 c_{p2}(T_2 - T) + \dot{Q} - \dot{W}_S$$
$$+ c_{A1} q_1 \left[\tilde{h}_{A1} - \tilde{h}_A + \frac{c_{B1}}{c_{A1}}(\tilde{h}_{B1} - \tilde{h}_B)\right]$$
$$+ c_{A2} q_2 \left[\tilde{h}_{A2} - \tilde{h}_A + \frac{c_{B2}}{c_{A2}}(\tilde{h}_{B2} - \tilde{h}_B)\right]. \tag{14.29}$$

The terms involving partial molar enthalpies appear to be related to the enthalpies of mixing, but some manipulation is required to obtain the correct forms. (Note that the enthalpies of mixing involve pure component enthalpies, which are absent in this equation.) Consider the term $c_{A1} q_1[\tilde{h}_{A1} - \tilde{h}_A + \frac{c_{B1}}{c_{A1}}(\tilde{h}_{B1} - \tilde{h}_B)]$. We can rewrite this as

$$c_{A1} q_1 \left[\tilde{h}_{A1} - \underline{h}_A + \frac{c_{B1}}{c_{A1}}(\tilde{h}_{B1} - \underline{h}_B)\right] - c_{A1} q_1 \left[\tilde{h}_A - \underline{h}_A + \frac{c_{B1}}{c_{A1}}(\tilde{h}_B - \underline{h}_B)\right]$$
$$= c_{A1} q_1 \Delta\underline{H}_{S1} - c_{A1} q_1 \left[\tilde{h}_A - \underline{h}_A + \frac{c_{B1}}{c_{A1}}(\tilde{h}_B - \tilde{h}_B)\right].$$

Performing a similar operation on the remaining term, together with a few further manipulations, we can write Equation 14.29 in the equivalent form

$$\rho V c_p \frac{dT}{dt} = \rho_1 q_1 c_{p1}(T_1 - T) + \rho_2 q_2 c_{p2}(T_2 - T) + c_{A1} q_1 \Delta\underline{H}_{S1}$$
$$+ c_{A2} q_2 \Delta\underline{H}_{S2} - (c_{A1} q_1 + c_{A2} q_2)\Delta\underline{H}_S + \dot{Q} - \dot{W}_S$$
$$+ \left[\frac{c_B}{c_A}(c_{A1} q_1 + c_{A2} q_2) - (c_{B1} q_1 + c_{B2} q_2)\right](\tilde{h}_B - \underline{h}_B). \tag{14.30}$$

These terms are all familiar – the three $\Delta\underline{H}_S$ terms represent the enthalpies of mixing of the two feedstreams and the vessel composition, for example – except for the final one, containing $\tilde{h}_B - \underline{h}_B$. This term is in fact identically zero for a steady-state composition ($dn_A/dt = dn_B/dt = 0$) or for "infinite dilution," where $\tilde{h}_B \approx \underline{h}_B$, as well as other special situations. In the general case we can derive the following relationship by differentiating Equation 14.12 with respect to n_B/n_A:

$$\tilde{h}_B - \underline{h}_B = \frac{\partial\Delta\underline{H}_S}{\partial(n_B/n_A)}. \tag{14.31}$$

As is evident from Figure 14.2, for small n_B/n_A this can be a very large quantity! The general form of the energy equation for a binary, nonreacting liquid system is then

$$\rho V c_p \frac{dT}{dt} = \rho_1 q_1 c_{p1}(T_1 - T) + \rho_2 q_2 c_{p2}(T_2 - T) + c_{A1} q_1 \Delta \underline{H}_{S1} + c_{A2} q_2 \Delta \underline{H}_{S2}$$

$$- (c_{A1} q_1 + c_{A2} q_2) \Delta \underline{H}_S + \left[\frac{c_B}{c_A} (c_{A1} q_1 + c_{A2} q_2) - (c_{B1} q_1 + c_{B2} q_2) \right]$$

$$\times \frac{\partial \Delta \underline{H}_S}{\partial (n_B/n_A)} + \dot{Q} - \dot{W}_S. \qquad (14.32)$$

This is our working equation. For some applications, especially design calculations at steady state, it is more convenient to evaluate enthalpies at the reference temperature T^o at which enthalpies of mixing are tabulated. The full transient equation* has many terms that vanish at steady state. We leave the conversion to that form as an exercise; at steady state the result is

$$T_{ss} = T^o + \beta \frac{c_{p1}}{c_p} (T_1 - T^o) + (1 - \beta) \frac{c_{p2}}{c_p} (T_2 - T^o)$$

$$+ \frac{x_A}{M_{wA} c_p} [-\Delta \underline{H}_S^o + \nu \Delta \underline{H}_{S1}^o + (1 - \nu) \Delta \underline{H}_{S2}^o] + \frac{\dot{Q} - \dot{W}_s}{\rho q c_p} \qquad (14.33)$$

where $\beta = \rho_1 q_1 / \rho q$, $\nu = q_1 c_{A1} / q c_A$, and x_A is the steady-state mass fraction of A in the tank. For computation it is useful to recall that $n_B/n_A = c_B/c_A = M_{wA}(1 - x_A)/M_{wB} x_A$. Note that for feedstreams that are pure there is a formal equivalence between Equation 14.32 and Equation 14.22b for semibatch mixing, except in the heat addition and shaft work terms.

EXAMPLE 14.5 Suppose we wish to mix equal mass flow rates of pure streams of water at $0°C$ (stream 1) and sulfuric acid at $-4°C$ in an insulated vessel. What is the temperature in the tank?

We assume that we are at steady state, in which case the tank temperature is constant and there are equal masses of the two components in the tank, and that the work of mixing can be neglected. We then have $\beta = 0.5$, $x_A = 0.5$, and $n_B/n_A = 98/18 = 5.4$. The relevant physical properties are given in Example 14.3, and we will assume again for simplicity that the heat capacities are constant. The enthalpies of mixing in the feedstreams are zero, so Equation 14.32 becomes identical to Equation 14.22b and we obtain a steady-state tank temperature $T_{ss} = 115°C$.

EXAMPLE 14.6 Now suppose that we wish to design a jacketed heat exchange system to maintain the temperature in the tank at $90°C$, using an aqueous stream at $T_{jf} = 20°C$ as the coolant. We will assume that the throughput to the mixing tank is 3 kg/min in both the acid and water streams. What is the required heat removal rate, and what are the design specifications of the heat transfer system?

The mass flow rate $\rho q = 6$ kg/min $= 0.1$ kg/s. It readily follows from inspection of Equation 14.32 that the rate of heat removal is simply $\rho q c_p$ multiplied by the difference between the adiabatic temperature computed in Example 14.5

---

* $\rho V c_p \frac{dT}{dt} = \rho_1 q_1 c_{p1}(T_1 - T^o) + \rho_2 q_2 c_{p2}(T_2 - T^o) - \left(\rho q c_p + M_{wA} c_{pA} \frac{dn_A}{dt} + M_{wB} c_{pB} \frac{dn_B}{dt} \right)(T - T^o) + c_{A1} q_1 \Delta \underline{H}_{S1}^o + c_{A2} q_2 \Delta \underline{H}_{S2}^o - (c_{A1} + c_{A2} q_2) \Delta \underline{H}_S^o + \left(\frac{c_B}{c_A} \frac{dn_A}{dt} - \frac{dn_B}{dt} \right) \frac{\partial \Delta H_S}{\partial (n_B/n_A)} - \dot{Q} - \dot{W}_S.$

($115°C$) and the desired temperature of $90°C$. We therefore obtain a required heat removal rate of 6,250 W. The design specifications for this heat removal rate were carried out in Example 13.2, where we used a heat transfer coefficient of 300 W/m^2 K for aqueous liquids in a steel tank. The minimum heat transfer area was determined to be 0.30 m^2 and the minimum coolant flow rate to be 1.26 kg of water per minute; the coolant flow rate for a heat transfer area of 1 m^2 would be 1.8 kg/min. The minimum area and flow rate would be the same if we were to use a cooling coil instead of a jacket, but the required cooling water flow rate for a heat transfer area of 1 m^2 was found in Example 13.3 to be 1.31 kg/min, just slightly above the minimum. In either case, the cooling water requirement is roughly half of the process water stream in the mixer. An important part of the design problem would be to minimize overall water usage and to optimize the temperatures of the various water streams in the process.

14.7 Acid-to-Water/Water-to-Acid

There are interesting features of the acid-to-water versus water-to-acid issue examined in Example 14.4 that are revealed by considering the transient temperature Equation 14.32. When the inlet streams contain only A in stream 1 (which we denote as stream A here) and only B in stream 2 (denoted as stream B), and there is no outflow, the equation for adiabatic mixing with negligible mixing work simplifies to

$$\rho V c_p \frac{dT}{dt} = \rho_A q_A c_{pA}(T_A - T) + \rho_B q_B c_{pB}(T_B - T) - \frac{\rho_A q_A}{M_{wA}} \Delta \underset{\sim}{H}_S$$
$$+ \left(\frac{n_B}{n_A} \frac{\rho_A q_A}{M_{wA}} - \frac{\rho_B q_B}{M_{wB}} \right) \frac{\partial \Delta \underset{\sim}{H}_S}{\partial (n_B/n_A)}. \tag{14.34}$$

The total mass ρV, the mixture heat capacity c_p, the molar ratio n_B/n_A, and the enthalpy of mixing $\Delta \underset{\sim}{H}_S$ all change with time as the vessel is filled, so this is a complex equation. It is possible to show that Equation 14.21 is an exact solution to this equation;[*] this is a remarkable result, but some of the physical behavior is best understood by looking directly at Equation 14.34. We consider again the two cases: Water (B) is added to acid (A) in the tank (Case I), and acid is added to water in the tank (Case II). For Case II, acid added to water, nearly all of the process occurs in the infinite dilution regime, where the enthalpy of mixing is essentially constant and the final term in Equation 14.34 vanishes. We thus have

$$\text{Case I: } \rho V c_p \frac{dT}{dt} = \rho_B q_B c_{pB}(T_B - T) - \frac{\rho_B q_B}{M_{wB}} \frac{\partial \Delta \underset{\sim}{H}_S}{\partial (n_B/n_A)}, \tag{14.35a}$$

$$\text{Case II: } \rho V c_p \frac{dT}{dt} = \rho_A q_A c_{pA}(T_A - T) - \frac{\rho_A q_A}{M_{wA}} \Delta \underset{\sim}{H}_S. \tag{14.35b}$$

---

[*] Simply differentiate $n_A \underset{\sim}{c}_{pA}(T_A - T) + n_B \underset{\sim}{c}_{pB}(T_B - T) = n_A \Delta \underset{\sim}{H}_S(n_B/n_A, T) + Q$ term by term with respect to t. The key step is noting that

$$\frac{d\Delta \underset{\sim}{H}_S}{dt} = \frac{\partial \Delta \underset{\sim}{H}_S}{\partial T} \frac{dT}{dt} + \frac{\partial \Delta \underset{\sim}{H}_S}{\partial (n_B/n_A)} \frac{d(n_B/n_A)}{dt} = \frac{\partial \Delta \underset{\sim}{H}_S}{\partial T} \frac{dT}{dt} + \frac{1}{n_A} \frac{\partial \Delta \underset{\sim}{H}_S}{\partial (n_B/n_A)} \left(\frac{dn_B}{dt} - \frac{n_B}{n_A} \frac{dn_A}{dt} \right).$$

The first term contributes to the heat capacity of the mixture (*cf.* Equation 4.14.)

Now, for Case II, the second term on the right-hand side of Equation 14.35b is always greater than the first term by at least a factor of three, so dT/dt is always positive and the approach to the final temperature is monotonic from below. Analysis of Case I requires a simple means of evaluating the second term, which we do by noting that the curve for H_2SO_4 in Figure 14.2 can be roughly approximated by the equation $\Delta H_S \sim -73,000[1 - \exp(-0.3n_B/n_A)]$, so $\partial \Delta H_S/\partial(n_B/n_A) \sim -21,900\exp(-0.3n_B/n_A)$. The second term is larger than the first by a factor of two when the two streams have mixed completely and $n_B/n_A = 5.4$, so the derivative is negative and the approach to the final state is from above. Hence, there must be a temperature overshoot at an intermediate composition, as shown in Figure 14.4.

It is also instructive to consider the initial rates of change of the temperatures in the two cases. The initial masses in the tank (ρV) and the mass flow rates (ρq) are the same. The temperature difference terms on the right-hand sides of Equations 14.35a and b are initially negligible, since the feed temperatures differ from the temperatures of the initial material in the tank by only $4°C$. The heat capacity at $t = 0$ for Case I is c_{pA}, since the starting material in the tank is pure A; similarly, the heat capacity at $t = 0$ for Case II is c_{pB}. $\partial \Delta H_S/\partial(n_B/n_A) \sim -21,900$ at $t = 0$. Hence,

$$\frac{dT_I/dt}{dT_{II}/dt}\bigg|_{t=0} \approx \frac{c_{pB}M_{wA}[\partial \Delta H_S/\partial(n_B/n_A)]_{n_B/n_A=0}}{c_{pA}M_{wB}[\Delta H_S]_{n_B/n_A=\infty}} = 4.9.$$

This substantially faster temperature rise when water is added to acid is evident in Figure 14.4. One point of interest is that the most important contribution to this phenomenon is the different heat capacities of the two liquids, which differ by a factor of three.

14.8 Concluding Remarks

The partial molar enthalpy is the most important concept introduced in this chapter, because it forms the basis for the analysis of thermal effects in all multicomponent systems, including reacting systems, which are addressed in the next chapter. It is quite common to find shortcut derivations that attempt to avoid the use of partial molar enthalpies. These shortcuts are invariably incorrect; they may lead to the correct working equations for special cases, especially those for which the correct equations are known, but such approaches always fail in complex situations. The chemical engineering literature is replete with incorrect energy balances obtained using shortcuts that give misleading and incorrect results. The approach developed here, illustrated with thermal effects in mixing, is often tedious, but it is straightforward and will always give the correct working equations.

It is important to emphasize once more that the treatment in this chapter is restricted to liquid-phase systems far from the critical point. Gaseous and multiphase systems are approached in the same manner, but the treatment is more complicated because of the need to address compressibility and its consequences.

Bibliographical Notes

The material in this chapter is covered in considerably more detail in chemical engineering thermodynamics courses, which typically comprise one or two core courses in the chemical engineering curriculum. All of the English-language textbooks in use at the time of writing contain the words *chemical*, *engineering*, and *thermodynamics* in the titles, usually in the order shown here.

The catalog of incorrect energy balances mentioned in the introduction is in

Denn, M. M, *Process Modeling*, Longman, London and Wiley, New York, 1986, ch. 5.

PROBLEMS

14.1 a. 10 L each of ethanol at –1.1°C (30°F), ethylene glycol at –6.67°C (20°F), and water at 26.7°C (80°F) are poured into a tank and mixed. What is the final temperature? Relevant physical property data are

	c_p (kJ/kg °C)	ρ (kg/m^3)
ethanol	2.17	790
ethylene glycol	2.30	1,130
water	4.18	1,000

b. The final temperature must be 21.1°C (70°F). Only the inlet temperature of the water can be adjusted. What should the water temperature be?

14.2 An insulated tank contains 10 kg of water at 20°C. A stream of 100 percent sulfuric acid at 20°C is added by accident at a rate of 1 kg/min. How long will it take for the temperature to rise to 100°C?

14.3 A continuous mixing operation is carried out by adding 0.3 kg/hr of calcium chloride crystals to a pure water stream flowing at 1 kg/hour.

a. If both streams are at 25°C when they enter the mixer, what is the temperature of the exit stream?

b. How much heat would have to be removed if the operation were to be carried out isothermally at 25°C?

The heat capacity of CaCl crystals is approximated by $c_p = 0.636 + 1.41 \times 10^{-4}T$ J/g K, where T is in K.

14.4 a. 5 kg/hr of sodium hydroxide and 10 kg/hr of water are to be mixed in an insulated tank. Both feedstreams are to be put into the mixer at the same temperature. What is a safe inlet temperature at which to mix the feeds?

b. We would like to carry out the mixing process in part (a) with the feedstreams at 20°C, and we do not wish the effluent temperature to exceed 80°C. Cooling water is available at 10°C, and we would like to use some or all of the cooling water to provide the 10 kg/hr water feedstream to the mixer. Is this process feasible using a jacketed vessel? Assume that the heat transfer coefficient h_T is of order 400 W/m^2 K.

15 Energy Balances for Reacting Systems

15.1 Introduction

The chemical reactor is the heart of most industrial processes, although the reactor may represent only about 10 percent of the total capital cost; this is because the output from the reactor defines everything else that must be done downstream, particularly the separation processes. Similarly, if we wish to think about cellular rather than industrial processes, it is the chemical reactions that enable the cell or the organ to carry out its essential functions. We saw in Chapters 7 through 9 how a chemical reactor is integrated into simple processes, and we explored economic issues such as the trade-off between capital and operating costs. That discussion was limited, however, because we assumed in every case that the rate constants were fixed numbers. In doing so, we ignored one of the "handles" that the chemical engineer – or, in the case of an organism, evolution – has available to promote efficiency. Chemical reaction rates are highly temperature dependent, and precise temperature control can be critical in both the design and functioning of a reactor.

Chemical reaction engineering is a broad subject, and it typically occupies at least one full course in an undergraduate chemical engineering curriculum. We introduce some basic ideas here for completeness, but we are only touching on one of the foundations of the chemical engineering profession. We restrict ourselves throughout this chapter to liquid systems, as before, in order to simplify some of the analysis while retaining the essential features, and we address only well-mixed reactor configurations.

15.2 Temperature Dependence of Reaction Rates

The temperature dependence of the reaction rate is most easily determined by performing a series of isothermal experiments, as discussed in Chapter 6, over the temperature range of interest. It is found in nearly all cases that a plot of the logarithm of the rate constant versus the reciprocal of the absolute temperature is linear, with

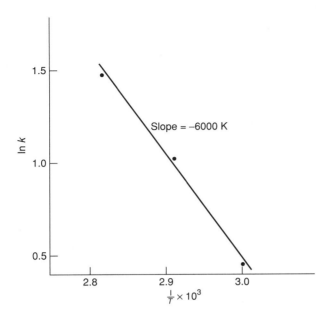

Figure 15.1. Rate constant in L/mol s as a function of reciprocal absolute temperature in K for the decomposition of sodium dithionate. Data of Rinker *et al.*, *Ind. Eng. Chem. Fundamentals*, **4**, 282 (1964).

a negative slope; that is, the reaction rate constant for a reaction of any order will usually follow the *Arrhenius* form,

$$k = k_o e^{-E/RT}, \tag{15.1}$$

where R is the ideal gas constant, 8.314 J/K mol, and E is known as the *activation energy*. A typical data set, for the decomposition of sodium dithionate, which is a second-order reaction with a rate constant of the form $r = kc_{S_2O_4^=}c_{H^+}$, is shown in Figure 15.1. The slope of ln k versus $1/T$ is approximately –6,000 K, so the value of E is approximately 50,000 J/mol. Values of E in the range 40,000 to 120,000 J/mol are typical. The Arrhenius form can be deduced from first principles, but theoretical predictions of the parameters k and E are generally unsatisfactory, and experimental measurement is usually required. It is important to keep in mind that the forward and reverse rates in a reversible reaction will each have a temperature dependence of Arrhenius form, but with different activation energies.

15.3 Heat of Reaction

Bonds are broken and reformed in any chemical reaction, entailing internal energy changes that are observed macroscopically in an insulated vessel by temperature increases or decreases. We know from the treatment in the preceding chapter that the partial molar enthalpies of the component species will enter the description of the reaction vessel, and we can anticipate that a combination of the component partial molar enthalpies will emerge in the final formulation, as it did with the heat of mixing in Chapter 14. This combination is the enthalpy change on reaction, commonly called the *heat of reaction* and denoted $\Delta \underset{\sim}{H}_R$, which is defined as follows:

Let the reactants be denoted as $A_1, A_2, A_3, \ldots$ and the products as $D_1, D_2, D_3, \ldots$ for the reaction $\alpha_1 A_1 + \alpha_2 A_2 + \alpha_3 A_3 + \cdots \rightleftharpoons \delta_1 D_1 + \delta_2 D_2 + \delta_3 D_3 + \cdots$. The $\{\alpha_i\}$ and $\{\delta_i\}$ are the stoichiometric coefficients; for example, in the reaction $H_2SO_4 + (C_2H_5)_2SO_4 \rightleftharpoons 2C_2H_5SO_4H$, which we studied in Example 6.7.1, we have $H_2SO_4 = A_1$, $(C_2H_5)_2SO_4 = A_2$, and $C_2H_5SO_4H = D_1$, with $\alpha_1 = \alpha_2 = 1$ and $\delta_1 = 2$. The heat of reaction is defined in terms of the partial molar enthalpies of the component species in a manner analogous to the definition of the heat of mixing:

$$\Delta \underline{H}_R \equiv \delta_1 \tilde{h}_{D_1} + \delta_2 \tilde{h}_{D_2} + \delta_3 \tilde{h}_{D_3} + \cdots - \alpha_1 \tilde{h}_{A_1} - \alpha_2 \tilde{h}_{A_2} - \alpha_3 \tilde{h}_{A_3} - \cdots. \quad (15.2)$$

Thus, the heat of reaction for the sulfuric acid-diethyl sulfate reaction is

$$\Delta \underline{H}_R \equiv 2\tilde{h}_{C_2H_5SO_4H} - \tilde{h}_{H_2SO_4} - \tilde{h}_{(C_2H_5)_2SO_4}.$$

Heats of reaction can be calculated from tabulated heats of formation. This is a topic that is often included in introductory courses in chemistry and may be familiar to some readers; it is included here as an appendix to this chapter so as not to disturb the development. Data are usually tabulated at a standard temperature, and they are often insensitive to temperature; derivation of the equation for the temperature dependence is left as an exercise.

15.4 The Batch Reactor – I

Heats of reaction are often measured in a batch reactor known as a *calorimeter*, which may be operated either isothermally or adiabatically, and the analysis of the batch reactor is, in any event, a good introduction to the role that the heat of reaction plays in process behavior. For definiteness we will consider the special case of the single reaction $A + B \rightleftharpoons \mu D$, which is the case considered in Section 6.7 (where mass action kinetics were assumed for the forward and reverse rates). Following the development in Chapter 6, we write the species mass balances in terms of the net reaction rate as

$$\frac{dn_A}{dt} = \frac{d}{dt}c_A V = -rV, \quad (15.3a)$$

$$\frac{dn_B}{dt} = \frac{d}{dt}c_B V = -rV, \quad (15.3b)$$

$$\frac{dn_D}{dt} = \frac{d}{dt}c_D V = +\mu rV. \quad (15.3c)$$

We are, of course, assuming that complete mixing has already occurred in the reaction vessel on a time scale that is fast relative to the rate at which the reaction progresses. The equations can be readily integrated (cf. Section 6.3) to give

$$n_A - n_{Ao} = n_B - n_{Bo} = -\frac{1}{\mu}(n_D - n_{Do}), \quad (15.4)$$

where n_{Ao}, n_{Bo}, and n_{Do} denote the respective number of moles of each component in the reactor at the start of the reaction.

The energy equation for this batch system is obtained by deleting the flow terms in Equation 12.5. Neglecting shaft work, the resulting equation is

$$\frac{dU}{dt} = \dot{Q} \approx \frac{dH}{dt}, \tag{15.5}$$

where we have replaced the derivative of internal energy with the derivative of enthalpy with little error for this liquid system. This equality between two time derivatives can be integrated immediately to obtain,

$$H(T, n_A, n_B, n_D) - H(T_o, n_{Ao}, n_{Bo}, n_{Do}) = Q. \tag{15.6}$$

T_o refers to the temperature in the reactor at the start of the reaction and Q is the total amount of heat added during the course of the reaction. The enthalpy is a function of the composition as well as the temperature, and we have noted the composition dependence explicitly in order to keep track of the proper values.

We need to calculate all enthalpies at the same temperature, as in the preceding chapter, and it is most convenient to use the reactor temperature as the reference. We will assume for convenience that the heat capacity of the starting mixture is independent of temperature; if this is not the case, temperature differences will simply be replaced by integrals. We can then write

$$H(T_o, n_{Ao}, n_{Bo}, n_{Do}) = H(T, n_{Ao}, n_{Bo}, n_{Do}) + \rho_o V_o c_{po}(T_o - T), \tag{15.7}$$

where ρ_o, V_o, and c_{po} refer to the density, volume, and heat capacity of the original mixture, respectively. (The product ρV is the total mass, which is a constant, so henceforth in this development we will drop the subscript o from the density and volume.) Equation 15.6 then becomes

$$H(T, n_A, n_B, n_D) - H(T, n_{Ao}, n_{Bo}, n_{Do}) = Q + \rho V c_{po}(T_o - T). \tag{15.8}$$

We now express each enthalpy in terms of the partial molar enthalpies:

$$H(T, n_A, n_B, n_D) = n_A \tilde{h}_A + n_B \tilde{h}_B + n_D \tilde{h}_D, \tag{15.9a}$$

$$H(T, n_{Ao}, n_{Bo}, n_{Do}) = n_{Ao} \tilde{h}_{Ao} + n_{Bo} \tilde{h}_{Bo} + n_{Do} \tilde{h}_{Do}. \tag{15.9b}$$

Here, $\tilde{h}_{io}$ refers to the partial molar enthalpy of species i in the initial mixture composition at temperature T. Finally, we use Equation 15.4 to solve for n_B and n_D in terms of n_A which, with some rearrangement, results in

$$(n_{Ao} - n_A)(\mu \tilde{h}_D - \tilde{h}_A - \tilde{h}_B) = Q + \rho V c_{po}(T_o - T) + \{n_{Ao}(\tilde{h}_{Ao} - \tilde{h}_A)$$
$$+ n_{Bo}(\tilde{h}_{Bo} - \tilde{h}_B) + n_{Do}(\tilde{h}_{Do} - \tilde{h}_D)\}. \tag{15.10}$$

The linear combination of partial molar enthalpies on the left is the heat of reaction. The three terms collected in the braces on the right are mixing terms that arise because of the changing composition. These mixing terms will usually be negligible compared to $\Delta \underset{\sim}{H}_R$ and we will neglect them[*]; if they are not negligible then there

---

[*] Consider, for example, the most common situation, in which $n_{Ao} = n_{Bo}$ and $n_{Do} = 0$. Then $n_A = n_B$ throughout the course of the reaction. If the product D forms a nearly ideal solution in A and B, then $\tilde{h}_A$ and $\tilde{h}_B$ are essentially unaffected by n_D and depend only on the molar ratio n_A/n_B, which is constant. In that case, $\tilde{h}_A = \tilde{h}_{Ao}$, $\tilde{h}_B = \tilde{h}_{Bo}$, and all three terms vanish.

is usually no convenient way to separate them from the measurement of $\Delta \underaccent{\tilde}{H}_R$ and they will introduce an error into the determination of the heat of reaction, but that problem is beyond the scope of our discussion here. Hence, the final equation for the heat of reaction in terms of the calorimeter experiment is

$$\Delta \underaccent{\tilde}{H}_R(T) = \frac{1}{n_{Ao} - n_A} [Q + \rho V c_{po}(T_o - T)]. \qquad (15.11)$$

The calorimeter is typically operated adiabatically ($Q = 0$). The reaction is run to completion and the final concentration (denoted $n_{A\infty}$) and temperature (denoted T_{ad}) are measured. The heat of reaction is then

$$\text{Adiabatic operation: } \Delta \underaccent{\tilde}{H}_R = \frac{\rho V c_{po}(T_o - T_{ad})}{n_{Ao} - n_{A\infty}}. \qquad (15.12a)$$

The calorimeter may also be operated with temperature control until completion of the reaction by adding or removing heat to ensure that the final and initial temperatures are the same. (Ideally we would do this isothermally, and we will refer to this mode as isothermal operation, but the result depends only on equal starting and ending temperatures.) In that case we obtain

$$\text{Isothermal operation: } \Delta \underaccent{\tilde}{H}_R = \frac{Q}{n_{Ao} - n_{A\infty}}. \qquad (15.12b)$$

An *exothermic reaction* is one in which there is an adiabatic temperature rise, or, equivalently, heat must be removed to maintain $T = T_o$. It follows from Equation 15.12a or b that $\Delta \underaccent{\tilde}{H}_R < 0$ for an exothermic reaction. (Recall that $n_{Ao} > n_{A\infty}$.) For an *endothermic reaction* the final adiabatic temperature is less than the initial temperature, or heat must be added to maintain the temperature. The notion that the heat of reaction is negative for an exothermic reaction is not an intuitive use of the English language, and we repeat this result for emphasis:

Exothermic: $\Delta \underaccent{\tilde}{H}_R < 0.$
Endothermic: $\Delta \underaccent{\tilde}{H}_R > 0.$

$\Delta \underaccent{\tilde}{H}_R$ is a function of temperature. Data are generally tabulated at a standard temperature denoted as T^o, which is frequently 25°C (298 K). If we were to repeat the steps leading to Equation 15.11, but with all enthalpies evaluated at the reference temperature T^o, we would obtain

$$\Delta \underaccent{\tilde}{H}_R(T^o) = \frac{1}{n_{Ao} - n_A} [Q + \rho V c_{po}(T_o - T^o) + \rho V c_p(T^o - T)]. \qquad (15.13)$$

c_p is the heat capacity of the final mixture. An equation for $\Delta \underaccent{\tilde}{H}_R$ at any temperature can be obtained from the calorimeter experiment by comparing Equations 15.11 and 15.13:

$$\Delta \underaccent{\tilde}{H}_R(T) = \Delta \underaccent{\tilde}{H}_R(T^o) + \frac{\rho V(c_p - c_{po})(T^o - T)}{n_{Ao} - n_A}. \qquad (15.14)$$

In many liquid systems there will be little difference between c_p and c_{po}, so $\Delta \underaccent{\tilde}{H}_R$ may often be taken as independent of temperature without serious error.

15.5 The Batch Reactor – II

It is convenient to derive the equation that describes the rate of change of the temperature in the batch reactor when there is a single reaction. Our starting point is again Equation 15.5, with U replaced by H for the liquid system. Since H depends on T, n_A, n_B, and n_D, we apply the chain rule to the derivative of enthalpy to obtain (*cf.* Equation 14.2)

$$\frac{dH}{dt} = \rho V c_p \frac{dT}{dt} + \tilde{h}_A \frac{dn_A}{dt} + \tilde{h}_B \frac{dn_B}{dt} + \tilde{h}_D \frac{dn_D}{dt} = \dot{Q}. \tag{15.15}$$

We use Equations 15.3a,b and c to replace the molar rates of change by the reaction rate, and we replace the linear combination of partial molar enthalpies by the heat of reaction using Equation 15.2:

$$\rho V c_p \frac{dT}{dt} = [\mu \tilde{h}_D - \tilde{h}_A - \tilde{h}_D] r V + \dot{Q} = (-\Delta \underset{\sim}{H}_R) r V + \dot{Q}. \tag{15.16}$$

Equation 15.16 must be solved simultaneously with the mass balance Equation 15.2 and the constitutive equation for the rate (e.g., mass-action kinetics and the Arrhenius temperature dependence).

It is instructive to make two assumptions:

(i) The heat capacity c_p is independent of composition and temperature.
(ii) The heat of reaction is independent of composition and temperature.

Together with Equation 15.3a, and noting that the total mass ρV is a constant, we can then write

$$\rho V c_p \frac{dT}{dt} - (\Delta \underset{\sim}{H}_R) \frac{dn_A}{dt} - \frac{dQ}{dt} = \frac{d}{dt} [\rho V c_p T - (\Delta \underset{\sim}{H}_R) n_A + Q] = 0, \tag{15.17}$$

which upon integration yields Equation 15.11. Hence, these two assumptions are sufficient to ensure that the partial molar enthalpy terms neglected in passing from Equation 15.10 to 15.11 are negligible. The assumption of a liquid heat capacity that is insensitive to composition is usually a reasonable one and is, in any event, subject to experimental verification. The concentration dependence of $\Delta \underset{\sim}{H}_R$ is difficult to obtain and rarely available, so it is common practice to assume a constant $\Delta \underset{\sim}{H}_R$ in the absence of better information.

The adiabatic batch reactor with a single reaction is an interesting special case. The temperature can be eliminated from the rate equation by writing Equation 15.11 with $Q = 0$ as

$$T = T_o + \frac{c_{Ao} - c_A}{\rho c_p} (-\Delta \underset{\sim}{H}_R). \tag{15.18}$$

The temperature-dependent term in the rate equation then becomes

$$\exp\left(\frac{-E}{RT}\right) = \exp\left(\frac{-E}{R} \left[\frac{1}{T_o + \dfrac{c_{Ao} - c_A}{\rho c_p}(-\Delta \underset{\sim}{H}_R)} \right]\right). \tag{15.19}$$

If the reaction is reversible, then the temperature-dependent terms in both the forward and reverse rates take the form of Equation 15.19, each with its own activation energy. The reaction rate, whatever the details of its form, is now a function of the single variable c_A; any dependence on the other component species can be related to c_A through Equation 15.4, while the temperature dependence is given in terms of c_A by Equation 15.19. If we assume that the volume is a constant then Equation 15.3a is simply of the form $dc_A/dt = -r(c_A)$, which can be expressed as a quadrature:

$$t = -\int_{c_{Ao}}^{c_A(t)} \frac{1}{r(c_A)} dc_A = \int_{c_A(t)}^{c_{Ao}} \frac{1}{r(c_A)} dc_A. \tag{15.20}$$

Closed-form integration is impossible because of the Arrhenius dependence of the temperature portion of the rate (although it can sometimes be approximated by functions that can be integrated in closed form), but the integration can be done numerically for any functional form of r using a method like the trapezoidal rule.

15.6 Continuous-Flow Stirred-Tank Reactor

The continuous-flow stirred-tank reactor (CFSTR) was treated in considerable detail in Chapters 6 and 7, where the basic mass balance equations were developed and design calculations were carried out for steady-state operation. The total design problem requires that the energy equation be taken into account, and we shall derive that equation here and consider some of its consequences.

The flow configuration is shown in Figure 15.2, where a jacket is included to maintain the reactor temperature at the desired value. In order to keep the algebra to a minimum we show the feedstreams as mixing just prior to entering the reactor, so there is a single inflow to the reactor. The flow in and out is maintained at the same volumetric flow rate q, and the standard assumptions are made regarding the density (Appendix 6A), so that the volume remains constant under these conditions. We again assume that there is a single reaction of the form $A + B \rightleftarrows \mu D$, for which

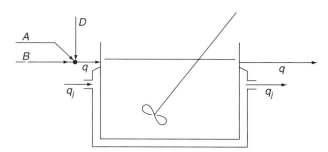

Figure 15.2. Schematic of the reaction $A + B \rightleftarrows \mu D$ in a jacketed CFSTR.

the equations of conservation of mass are

$$\frac{dn_A}{dt} = V\frac{dc_A}{dt} = q\left(c_{Af} - c_A\right) - rV, \tag{15.21a}$$

$$\frac{dn_B}{dt} = V\frac{dc_B}{dt} = q\left(c_{Bf} - c_B\right) - rV, \tag{15.21b}$$

$$\frac{dn_D}{dt} = V\frac{dc_D}{dt} = q\left(c_{Df} - c_D\right) + \mu rV. \tag{15.21c}$$

The equation of conservation of energy applied to the reactor is

$$\frac{dU}{dt} \approx \frac{dH}{dt} = \rho_f q h_f(T_f) - \rho q h(T) + \dot{Q} - \dot{W}. \tag{15.22}$$

Here we have again taken the rate of change of the internal energy to be equal to the rate of change of the enthalpy, which is a valid approximation for this liquid system. In what follows we will set the shaft work to zero; shaft work is an additive term that can always be put back in if appropriate, but it is typically unimportant for liquid systems for which perfect mixing can be achieved. The enthalpy of the feedstream is evaluated at the reactor temperature through Equation 12.18b:

$$h_f(T_f) = h_f(T) + c_{pf}(T_f - T). \tag{15.23}$$

In addition, we have the following identities (Equation 14.7):

$$\rho_f h_f(T) = c_{Af}\tilde{h}_{Af}(T) + c_{Bf}\tilde{h}_{Bf}(T) + c_{Df}\tilde{h}_{Df}(T), \tag{15.24a}$$

$$\rho h(T) = c_A\tilde{h}_A(T) + c_B\tilde{h}_B(T) + c_D\tilde{h}_D(T). \tag{15.24b}$$

With Equation 14.2, which is an application of the chain rule, we write

$$\frac{dH}{dt} = \rho V c_p \frac{dT}{dt} + \tilde{h}_A\frac{dn_A}{dt} + \tilde{h}_B\frac{dn_B}{dt} + \tilde{h}_D\frac{dn_D}{dt}. \tag{15.25}$$

Combining Equations 15.21 through 15.25 and eliminating terms that appear on both sides of the equation (compare the steps leading to Equation 14.28) then gives

$$\rho V c_p \frac{dT}{dt} = \rho_f q c_{pf}(T_f - T) - \left(\mu\tilde{h}_D - \tilde{h}_A - \tilde{h}_B\right)rV$$
$$+ q\left[c_{Af}(\tilde{h}_{Af} - \tilde{h}_A) + c_{Bf}(\tilde{h}_{Bf} - \tilde{h}_B) + c_{Df}(\tilde{h}_{Df} - \tilde{h}_D)\right] + \dot{Q}. \tag{15.26}$$

The term $\mu\tilde{h}_D - \tilde{h}_A - \tilde{h}_B$ is simply the heat of reaction, $\Delta\underline{H}_R$, whereas the other terms involving enthalpy differences reflect the enthalpy change upon mixing the feedstream with the reactor contents. As discussed in Sections 15.3 and 15.4, these terms will usually be small compared to $\Delta\underline{H}_R$ and will vanish identically in an ideal solution, and we shall neglect them here. Thus, the energy equation with a single reaction becomes

$$\rho V c_p \frac{dT}{dt} = \rho q c_{pf}(T_f - T) + (-\Delta\underline{H}_R)rV + \dot{Q}. \tag{15.27}$$

If the reactor is cooled or heated by a jacket, as shown in Figure 15.2, then the equation for the jacket temperature can be written (cf. Section 13.3)

$$\rho_j V_j c_{pj} \frac{dT_j}{dt} = \rho_j q_j c_{pj}(T_{jf} - T_j) - \dot{Q}, \tag{15.28}$$

where the subscript j refers to the jacket and we have made use of the fact that $Q_j = -\dot{Q}$. The rate of heat transfer is expressed in terms of the heat transfer area and heat transfer coefficient as

$$Q = -h_T a(T - T_j). \tag{15.29}$$

The equations for the reactor and jacket are then, finally,

$$\rho V c_p \frac{dT}{dt} = \rho q c_{pf}(T_f - T) + (-\Delta \underline{H}_R) r V - h_T a(T - T_j), \tag{15.30}$$

$$\rho_j V_j c_{pj} \frac{dT_j}{dt} = \rho_j q_j c_{pj}(T_{jf} - T_j) + h_T a(T - T_j). \tag{15.31}$$

The design equations for a CFSTR with a single reaction are obtained by setting the time derivatives in Equations 15.21, 15.30, and 15.31 to zero and incorporating the reaction rate constitutive equation. Note that the design problem decomposes into two parts: At any given temperature the equations are those treated in Chapter 7, and the full discussion there is relevant. We must, however, specify the reactor temperature before any design is complete, and this adds an additional set of considerations to those discussed in Chapter 7, including not only the economics but also, as we shall see, the core issue of reactor operability.

15.7 Steady-State CFSTR

The algebraic equations for the steady-state CFSTR have an interesting mathematical structure that has a significant impact on the design and actual operation of the reactor. This structure manifests itself nicely with a minimum of mathematical complexity by considering the irreversible first-order reaction A → D, for which the rate is

$$r = kc_A = k_o e^{-E/RT} c_A. \tag{15.32}$$

We define the following parameters:

$$\theta = \frac{V}{q}, \quad J = \frac{-\Delta \underline{H}_R}{\rho c_{pf}}, \quad K = \frac{h_T a}{\rho_j q_j c_{pj}}, \quad u_T = \frac{\rho_j q_j c_{pj}}{\rho q c_{pf}} \frac{h_T a}{h_T a + \rho_j q_j c_{pj}}. \tag{15.33a,b,c,d}$$

θ is the reactor residence time, while the other parameters reflect relative thermal effects. At steady state ($d/dt = 0$), Equations 15.21a, 15.30, and 15.31 then become

$$0 = c_{Af} - c_A - k_o \theta e^{-E/RT} c_A, \tag{15.34a}$$

$$0 = T_f - T + J k_o \theta e^{-E/RT} c_A - u_T(1 + K)(T - T_j), \tag{15.34b}$$

$$0 = T_{jf} - T_j + K(T - T_j). \tag{15.34c}$$

We can solve Equation 15.34c for T_j and substitute the result into Equation 15.34b so that we are left with two coupled equations for the two variables c_A and T. It is then convenient to multiply Equation 15.34a by J and add the result to

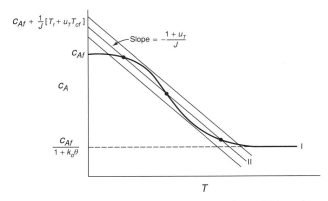

Figure 15.3. c_A as a function of T. Line I is Equation 15.35a, whereas Line II is Equation 15.35b. The intersection defines possible steady-state solutions.

Equation 15.34b, which will remove the reaction rate term from the sum. We then solve the two equations for c_A explicitly in terms of T, as follows:

$$\text{I: } c_A = \frac{c_{Af}}{1 + k_o\theta e^{-E/RT}}, \tag{15.35a}$$

$$\text{II: } c_A = c_{Af} + \frac{1}{J}(T_f + u_T T_{jf}) - \frac{1 + u_T}{J}T. \tag{15.35b}$$

We could take one more step and eliminate c_A between these two equations to obtain a single nonlinear algebraic equation for T, but the structure and interesting physical behavior is best revealed in this form.

We now plot c_A as a function of T according to each of the two equations, 15.35a and 15.35b. Equation 15.35a has the form shown as Line I in Figure 15.3. As $T \to 0$, $c_A \to 0$ with a zero slope, while as $T \to \infty$, T approaches $c_{Af}/(1 + k_o\theta)$ asymptotically. Equation 15.35b, on the other hand, is a straight line with slope $-(1 + u_T)/J$, as shown as Line II in Figure 15.3. The intersection of Lines I and II defines the pair (c_A, T) that provides the simultaneous solution to the reactor equations. There can be as many as three different solutions to these equations, depending on the values of the parameters. The jacket is not required for this behavior, which is also possible with an adiabatic reactor ($u_T = 0$).

The possibility of multiple solutions to the nonlinear steady-state equations is not surprising. Indeed, we have already seen such behavior in Chapter 8; the bioreactor with organism recycle could operate in either of two states, for example, one of which is "washout," and the approach of plotting two curves and looking for multiple intersections was used in the analysis of the CO oxidation reaction. Here, the behavior is somewhat more interesting than that found for the nonlinear systems in Chapter 8. It is possible to show, by considering the dynamical response, that the intermediate steady state is not a feasible solution, but the other two are; that is, within a parameter range such that the curves have multiple intersections, the reactor might operate either at a high-temperature, high-conversion state or at a low-temperature, low-conversion state. Figure 15.4 shows experimental data for

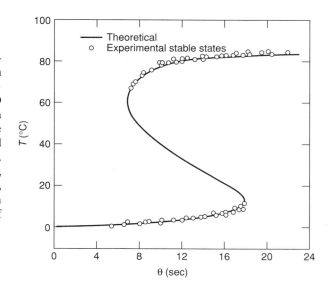

Figure 15.4. Experimental steady states for the reaction $2Na_2S_2O_3 + 4H_2O_2 \rightarrow Na_2S_3O_6 + Na_2S_2O_4 + 4H_2O$ in an adiabatic CFSTR as a function of residence time. The theoretical curve was computed from batch data. Data of S. A. Vejtasa and R. A. Schmitz, *AIChE Journal*, **16**, 410 (1970), reproduced with permission of the American Institute of Chemical Engineers.

just such a case. Here, the reaction $2Na_2S_2O_3 + 4H_2O_2 \rightarrow Na_2S_3O_6 + Na_2S_2O_4 + 4H_2O$ was carried out first in a batch reactor in order to obtain the kinetic data, and then in an adiabatic CFSTR. The steady-state temperature in the CFSTR is plotted as a function of the residence time θ in the figure. The solid line is the computed curve using the batch rate data. For residence times in the range of approximately 7 to 18 seconds there are three possible steady-state temperatures; two were found experimentally, one at high temperature and one at low. Note that, if we are operating the reactor at a sequence of steady states with gradually increasing residence times, there will be a sudden jump from the lower branch to the upper at the point where the curve turns back; there is hysteresis in the system, and the downward jump from the upper branch to the lower with decreasing residence time will occur at a different point. (Students who have studied combustion reactions in physical chemistry will recognize the analogy to the ignition-extinction phenomenon.)

The phenomenon of multiplicity is a real one that arises in many exothermic systems, and it has a substantial impact on reactor design and operation. The dynamical response of the system is critical, for it determines the steady state that will actually be attained for given initial and operating conditions. There is a danger that upsets will drive the system from the design state to another, unwanted steady state, perhaps accompanied by large temperature excursions that can damage the system or cause other safety problems, so integration of the design of the control system is a critical component of the overall design. The dynamics of a reactor with a single chemical reaction can be analyzed in a relatively straightforward way without any knowledge of differential equations, as shown in Appendix 15.B. The dynamics of systems with multiple reactions requires the treatment of families of nonlinear differential equations and is beyond the scope of our introductory treatment.

15.8 Design of CFSTR Systems

We found in Chapter 7 that the optimal design of a continuous-flow stirred-tank reactor required considering the effect of the reactor on the economics of the entire process. In that discussion we implicitly assumed that the reactor temperature was known, and nothing was said about the best operating temperature. Although a complete discussion of that problem is well outside the scope of this introductory text, we can present some elements of the logic needed to determine the optimal operating temperature for a stirred-tank reactor system.

Equations 15.34a, b, and c are the basic design equations for the single irreversible reaction A $\rightarrow$ products. Batch data provide us with values of k_o, E, and ΔH_R, and we will assume that the inlet conditions c_{Af}, T_f, and T_{jf} are specified by process requirements (raw material availability, stream source, etc.) and are not subject to our control. We can therefore control the operating temperature only through the conversion and the design of the heat exchange system. The simplest design will have no heat exchange, and this case should always be considered first. The strategy is therefore clear: We design for adiabatic operation and then consider the economic consequences of either heating or cooling the reactor. A cooling system can usually be justified when the adiabatic design leads to temperatures that are unattainable with reasonable materials of construction, when adiabatic design leads to excessively high working pressures, or when adiabatic design causes a loss of production and the possible need for additional separation processes because of the production of unwanted byproducts at high temperatures. A heating system to raise the reactor temperature and increase the reaction rate can only be justified if its capital and operating costs are less than the additional capital and operating costs of the larger reactor that would be required for an adiabatic design. The same comments would apply to a preheater for the feedstreams if feed temperatures are allowed to vary.

For more complex reaction systems, such as those considered in Section 7.4, the effect of temperature on the product distribution, as well as on the reactor size, must be considered. Some insight into the general problem can be obtained by examining the specific case of the product distributions for R, S, and T defined by Equations 7.19a, b, c, and d. In these equations the temperature dependence is contained in the ratios of the rate constants, which will be of the form

$$\frac{k_2}{k_1} = \frac{k_{20}}{k_{10}} \exp\left(\frac{E_1 - E_2}{RT}\right), \quad \frac{k_3}{k_1} = \frac{k_{30}}{k_{10}} \exp\left(\frac{E_1 - E_3}{RT}\right).$$

The product distribution at any value of x_A will be affected by the operating temperature if E_1, E_2, and E_3 are unequal. In that case an exchanger that either heats or cools the reactor may be justified in order to obtain a more favorable mix of R, S, and T. If $E_1 = E_2 = E_3$, as is the case for the ethylene glycol reaction, the operating temperature has no effect on the product distribution and the optimal temperature may be found as discussed for the single reaction.

15.9 Concluding Remarks

The development of the heat of reaction for reacting systems is analogous to the development of the heat of mixing in Chapter 14, which also involves a linear combination of partial molar enthalpies. The reader who understands these developments will have no difficulty going on to treat more complex physical situations in multicomponent systems, including phase change and multiphase processing. The heats of mixing and reaction have the appearance of generation terms in a balance equation, and it is not uncommon to see "derivations" that include "heat generated by mixing" or "heat generated by reaction" as additive terms in the energy balance. But there is no generation of energy in a nonrelativistic system, and attempting nonphysical shortcuts that bypass the treatment in terms of partial molar enthalpies is a surefire way to get the equations wrong.

The design and operation of chemical reactors is one of the foundations of chemical engineering, regardless of whether the application is to one of the traditional processing industries or to nontraditional applications such as semiconductor processing or cell cultivation for artificial organs. An introductory chapter like this one can only touch on the subject; indeed, in order to go as far as we have with chemical reaction engineering in this text, it has been necessary to limit ourselves to liquid-phase systems, very elementary reaction schemes, and the most restrictive class of reaction vessel configurations. Much more will be addressed in courses on reaction engineering and process design, as well as elective courses focusing on specific applications where chemical reaction is an important component.

It is important to close this chapter with a reminder. By restricting ourselves to liquid-phase systems far from the critical point we have been able to make the simplification that pressure is not an important variable and that the pressure dependence of the enthalpy can be ignored. This, in turn, has enabled us to equate the rates of change of internal energy and enthalpy, which has simplified things immeasurably. This substitution cannot be made in gas-phase systems, or in liquid-phase systems in the neighborhood of the critical point; it is internal energy, not enthalpy, whose change is monitored in the conservation equation, and the liquid system is a simplified special case.

Bibliographical Notes

The physical property data sources given at the end of Chapter 12 are relevant here. There are many textbooks for first courses in reaction engineering with titles that include the words *kinetics* and *reaction engineering,* and the material in this chapter is covered in most. As noted in Chapter 14, there is a discussion of common errors in writing energy balances for reacting systems in Chapter 5 of

Denn, M. M., *Process Modeling*, Longman, London and Wiley, New York, 1986.

PROBLEMS

15.1. (Appendix 15A) Find the heat of reaction $\Delta \underline{H}_R$ for each of the following reactions using both heat of formation and heat of combustion data, as available.

(i) $C_6H_6 + HNO_3 \rightarrow C_6H_5NO_2 + H_2O$

(ii) $C_2H_5OH + CH_3COOH \rightarrow CH_3C\underset{OC_2H_5}{\overset{O}{\diagup\diagdown}} + H_2O$

(iii) $CH_3-C\underset{Cl}{\overset{O}{\diagup\diagdown}} + H_2O \rightarrow CH_3-C\underset{OH}{\overset{O}{\diagup\diagdown}} + HCl$

(iv) $\begin{array}{c}CH_3-C\overset{O}{\diagup}\\[-2pt] \diagdown O\\ CH_3-C\underset{O}{\diagdown\!\!\!\diagdown}\end{array} + H_2O \rightarrow 2CH_3-C\underset{OH}{\overset{O}{\diagup\diagdown}}$

(v) $C-C\underset{OH}{\overset{O}{\diagup\diagdown}} + NH_4OH(aq) \rightarrow CH_3-C\underset{NH_2}{\overset{O}{\diagup\diagdown}} + 2H_2O$

(vi) $C_6H_6 + CH_3OH \rightarrow C_6H_5CH_3 + H_2O$

$C_6H_5CH_3 + CH_3OH \rightarrow C_6H_4(CH_3)_2 + H_2O$

$C_6H_4(CH_3)_2 + CH_3OH \rightarrow C_6H_3(CH_3)_3 + H_2O$

(vii) $CH_3CN + C_2H_5OH + H_2O \rightarrow CH_3C\underset{OC_2H_5}{\overset{O}{\diagup\diagdown}} + NH_4OH \; (aq)$

(viii) $H-\overset{\overset{\displaystyle H}{|}}{\underset{}{C}}\!\!-\!\!\overset{\overset{\displaystyle H}{|}}{\underset{O}{C}}\!\!-\!\!H + H_2O \rightarrow H-\overset{\overset{\displaystyle H}{|}}{\underset{\underset{\displaystyle OH}{|}}{C}}\!-\!\overset{\overset{\displaystyle H}{|}}{\underset{\underset{\displaystyle OH}{|}}{C}}\!-\!H$

(ix) $CH_2{=}C{=}O + CH_3C\underset{OH}{\overset{O}{\diagup\diagdown}} \rightarrow \begin{array}{c}CH_3-C\overset{O}{\diagup}\\[-2pt] \diagdown O\\ CH_3-C\underset{O}{\diagdown\!\!\!\diagdown}\end{array} \xrightarrow{C_2H_5OH} CH_3C\underset{OH}{\overset{O}{\diagup\diagdown}} + CH_3C\underset{OC_2H_5}{\overset{O}{\diagup\diagdown}}$

15.2. a. Derive an equation for the heat of reaction at any temperature if you are given the heat of reaction $\Delta \underline{H}_R(T^o)$ at a standard temperature T^o. You may assume an ideal solution.

b. Find the temperature dependence of the heat of reaction for reaction (viii) in Problem 15.1. (You will need to search for some data to do this.)

15.3. The following specific reaction rate constant data were reported by Smith [*J. Phys. & Coll. Chem.*, **59**, 367 (1947)] for the reaction urea + formaldehyde → monomethylolurea:

T, °C	30	40	50	60
k, L/mol s	5.5×10^{-5}	11.8×10^{-5}	24.5×10^{-5}	50.1×10^{-5}

Check the validity of the Arrhenius relation and determine the activation energy.

15.4. For the formation of glucose from cellulose discussed in Problem 6.12, the rate constant k_1 shows the following temperature dependence at an HCl concentration of 0.055 mol/L:

T, °C	160	170	180	190
k_1, min^{-1}	0.00203	0.00568	0.0190	0.0627

Check the validity of the Arrhenius relation and determine the activation energy.

15.5. A common rule of thumb states that the reaction rate doubles for each 10°C rise in temperature. Under what conditions is this rule of thumb valid?

15.6. The irreversible exothermic reaction A + B → nD is to be carried out isothermally in a batch reactor. The rate is described by mass action kinetics.

a. What is the rate at which heat must be removed from the reactor in order to maintain a constant temperature?

b. If the reactor is cooled with a jacket, how must the flow rate in the jacket be controlled in order to maintain isothermal operation? Assume that the cooling liquid is available at a fixed temperature T_{jf}.

15.7. The irreversible first-order reaction A → R is to be carried out in an adiabatic continuous flow stirred-tank reactor. The molecular weight of A is 18, and you may take all physical properties to be those of water and constant. The reaction rate $k = k_o \exp(-E/RT)$, with $k_o = 10^9$ min^{-1} and $E/R = 9{,}250$ K. $\Delta \tilde{H}_R$ −88.3 kJ/mol. The feed enters at 2.83 m^3/hr and 37.8°C and contains 0.444 kg/m^3 of A; 80 percent of the A is to be converted. Find the operating temperature and the volume of the reactor.

15.8. The catalytic gas phase oxidation of naphthalene to form phthalic anhydride is thought to take place as follows:

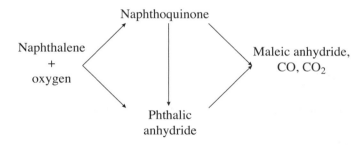

The reaction has been studied by DeMaria, Longfield, and Butler [*Ind. Eng. Chem.*, **53**, 259 (1961)] in a fluidized bed reactor with two catalysts, designated A and B. A surprisingly good way to model a fluidized bed reactor is to assume that it behaves like a CFSTR, using the same equations as for a liquid-phase system. When oxygen is in excess, which is always the case in commercial operation, the reaction is well described by the simple consecutive scheme naphthalene (A) → phthalic anhydride (M) → undesired combustion products (S), where both reactions are pseudo-first-order and irreversible.

Let $k_1 = k_{10} \exp(-E_1/RT)$ be the first-order rate constant for A → M and $k_2 = k_{20} \exp(-E_2/RT)$ be the first-order rate constant for M → S. The following data are available:

	Catalyst A	Catalyst B
E_1/R	21,200 K	10,300 K
E_2/R	10,100 K	23,100 K
k_1	$2.6 \times 10^{-4}\,\text{s}^{-1}$ at 224°C	$13\,\text{s}^{-1}$ at 497°C
k_2	$= k_1$ at 224°C	$= k_1$ at 497°C

a. Compute the selectivity $s = c_R/(c_{Af} - c_A)$ and the yield $y = c_R/c_{Af}$ in a reactor with a residence $\theta = V/q$ of one second (1s) and plot the results for a range of temperatures between 224 and 500°C.

b. Select the reactor operating temperature at the point where you estimate that there is a maximum in the yield curve. Pick two residence times that are different from 1s and see if you can improve the yield. Which catalyst is more efficient?

c. How does the selectivity affect the reactor design? Are there differences between catalysts A and B?

d. How would you select the optimal residence time for the phthalic anhydride reaction? (*cf.* Problem 7.5.)

Appendix 15A: Calculation of Heats of Reaction from Tabular Data

It is often possible to compute heats of reaction from tabular data. One such procedure is outlined here, together with the assumptions involved. Table 15A.1 shows the *heats of formation* of selected compounds at 25°C; the heat of formation of compound i, denoted ΔH_{Fi}, is the heat of reaction when a compound is formed from its elements in their natural states by the (possibly nonexistent) reaction Elements → Compound i. The relation that we will derive is

$$\Delta H_R = \delta_1 \Delta H_{FD_1} + \delta_2 \Delta H_{FD_2} + \cdots - \alpha_1 \Delta H_{FA_1} - \alpha_1 \Delta H_{FA_2} - \cdots, \qquad (15A.1)$$

where $\delta_1, \delta_2, \ldots, \alpha_1, \alpha_2, \ldots$ are the stoichiometric coefficients in the reaction

$$\alpha_1 A_1 + \alpha_2 A_2 + \alpha_3 A_3 + \cdots \rightleftarrows \delta_1 D_1 + \delta_2 D_2 + \delta_3 D_3 + \cdots. \qquad (15A.2)$$

The chemical equation for the formation of one mole of species i from the elements is

$$\varepsilon_i^1 E_1 + \varepsilon_i^2 E_2 + \varepsilon_i^3 E_3 + \cdots \rightarrow i. \qquad (15A.3)$$

Table 15A.1. *Heats of formation of selected*
compounds at 25°C.

	$\Delta \underset{\sim}{H}_F$, kJ/mol
n-pentane, C_5H_{12}	-173.5
n-hexane, C_6H_{14}	-198.7
n-octane, C_8H_{18}	-250.1
2,3-dimethyl butane, C_6H_{14}	-207.4
benzene, C_6H_6	$+49.1$
toluene, C_7H_8	$+12.4$
ethanol, C_2H_6O	-277.6
ethylene glycol, $C_2H_6O_2$	-460.0
ethylene oxide (liquid), $(CH_2)_2O$	$-77.6*$
HNO_3	-174.1
H_2O	-285.8
H_2SO_4	-811.3
KNO_3	-494.0
KOH	-428.8

* Ethylene oxide, often tabulated under the official IUPAC
name *oxirane,* is a gas at 25°C, so this is an artificial value that
includes the enthalpy change associated with the phase change.

Here, i refers to $A_1, A_2, A_3, \ldots, D_1, D_2, D_3$. $E_1, E_2, E_3, \ldots$ are the elements in their
natural states, and ε_i^k is the stoichiometric coefficient of element k in the reaction to
form compound i. The heat of formation is then

$$\Delta \underset{\sim}{H}_{Fi} = h_j - \varepsilon_i^1 \Delta \underset{\sim}{H}_{E_1} - \varepsilon_i^2 \Delta \underset{\sim}{H}_{E_2} - \varepsilon_i^3 \Delta \underset{\sim}{H}_{E_3} - \cdots. \qquad (15A.4)$$

We assume that there are no mixing effects, and pure component enthalpies are used.
We will first assume ideal behavior in the reaction mixture for Equation 15A.2, so
$\tilde{h}_i = \underset{\sim}{h}_i$. Then combining Equation 15.2 for $\Delta \underset{\sim}{H}_R$ and Equation 15A.4 for $\Delta \underset{\sim}{H}_F$ we
obtain

$$\begin{aligned}
\Delta \underset{\sim}{H}_R &= \delta_1 \tilde{h}_{D1} + \delta_2 \tilde{h}_{D2} + \delta_3 \tilde{h}_{D3} + \cdots - \alpha_1 \tilde{h}_{A1} - \alpha_2 \tilde{h}_{A2} - \alpha_3 \tilde{h}_{A3} - \cdots \\
&= \delta_1 \underset{\sim}{h}_{D1} + \delta_2 \underset{\sim}{h}_{D2} + \delta_3 \underset{\sim}{h}_{D3} + \cdots - \alpha_1 \underset{\sim}{h}_{A1} - \alpha_2 \underset{\sim}{h}_{A1} - \alpha_3 \underset{\sim}{h}_{A1} - \cdots \\
&= \delta_1 \left[\Delta \underset{\sim}{H}_{FD_1} + \varepsilon_{D_1}^1 \underset{\sim}{h}_{E_1} + \varepsilon_{D_1}^2 \underset{\sim}{h}_{E_2} + \cdots \right] \\
&\quad + \delta_2 \left[\Delta \underset{\sim}{H}_{FD_2} + \varepsilon_{D2}^1 \underset{\sim}{h}_{E_1} + \varepsilon_{D2}^2 \underset{\sim}{h}_{E_2} + \cdots \right] + \cdots \\
&\quad - \alpha_1 \left[\Delta \underset{\sim}{H}_{FA_1} + \varepsilon_{A_1}^1 \underset{\sim}{h}_{E_1} + \varepsilon_{A_1}^2 \underset{\sim}{h}_{E_2} + \cdots \right] \\
&\quad - \alpha_2 \left[\Delta \underset{\sim}{H}_{FA_2} + \varepsilon_{A2}^1 \underset{\sim}{h}_{E_1} + \varepsilon_{A2}^2 \underset{\sim}{h}_{E_2} + \cdots \right] - \cdots. \qquad (15A.5)
\end{aligned}$$

Consider the terms multiplying $\underset{\sim}{h}_{E_1}$: $\underset{\sim}{h}_{E_1} \left[\delta_1 \varepsilon_{D_1}^1 + \delta_2 \varepsilon_{D_2}^1 + \cdots - \alpha_1 \varepsilon_{A_1}^1 - \alpha_2 \varepsilon_{A_2}^1 - \cdots \right]$.
The sum $\delta_1 \varepsilon_{D_1}^1 + \delta_2 \varepsilon_{D_2}^1 + \cdots$ represents the total number of atoms of element 1
on the right-hand side of Equation 15A.2, whereas the sum $\alpha_1 \varepsilon_{A_1}^1 + \alpha_2 \varepsilon_{A_2}^1 + \cdots$
represents the total number of atoms of element 1 on the left. These sums must be
equal for the chemical equation to balance, so the coefficient of $\underset{\sim}{h}_{E_1}$ is simply equal to
zero. Similarly, the coefficients of $\underset{\sim}{h}_{E2}, \underset{\sim}{h}_{E3}, \ldots$ also equal zero, and Equation 15A.5
reduces to Equation 15A.1, where the heat of reaction is expressed in terms of the
heats of formation. The enthalpy of an element in its natural state will always drop

out of any formulation, so we could simplify the algebra by adopting the convention that *the enthalpy of an element in its natural state at the reference temperature is zero.* This convention is universally applied, and the reason that it works is evident from the development here.

EXAMPLE 15A.1 Compute ΔH_R for the reaction between ethylene oxide (A_1) and water (A_2) to form ethylene glycol (D_1):

$$H_2C{-}CH_2 + H_2O \rightarrow H_2C{-}CH_2.$$
$$\begin{array}{cc} \backslash\ / & \quad\quad | \quad | \\ O & \quad\quad HO \quad OH \end{array}$$

The heats of formation from Table 15A.1 are as follows: ethylene glycol, -460.0; ethylene oxide, -77.6; water, -285.8. Thus, $\Delta H_R = (-460.0) - (-77.6) - (-285.8) = -96.6$ kJ/mol. The reaction is exothermic, since $\Delta H_R < 0$.

EXAMPLE 15A.2 Aqueous solutions of 6.25 mole percent HNO_3 (1 mole of HNO_3 to 15 moles H_2O) and 6.25 mole percent KOH react to form a solution of KNO_3. Estimate the heat of reaction.

The reaction in solution is $HNO_3 + KOH \rightarrow KNO_3 + H_2O$. Hence, there will be 31 moles of H_2O for each mole of KNO_3 in the product stream. Heats of formation of pure liquid HNO_3 and H_2O and crystalline KOH and KNO_3 are given in Table 15A.1 as follows: HNO_3 liquid, -174.1; KOH crystal, -428.8; KNO_3 crystal, -494.0; H_2O liquid, -285.8. The reaction takes place in solution, so the relevant heat of formation is that of the compound in solution, for which we need to use the heats of solution, which are given in Figure 14.2. The potassium hydroxide solution is at infinite dilution, and the nitric acid and final salt solution are nearly so. The heats of solution from the figure (which are accurate to no more than two significant figures) are HNO_3, -31; KOH, -53; KNO_3, $+31$. There should be no further solution effects when the dilute acid and base are mixed, so the heat of formation in the reaction solution may be approximated by the sum of the heat of formation of the pure material and the heat of solution: HNO_3 solution, $-174 - 31 = -205$; KOH solution, $-429 - 53 = -482$; KNO_3 solution, $-494 + 31 = -463$; H_2O liquid, -286. Then $\Delta H_R = (-463) + (-286) - (-205) - (-482) = -62$ kJ/mole. The reaction is exothermic.

Heats of formation can be obtained for hydrocarbons from the tabulated *heat of combustion*, which is the heat of reaction for the exothermic reaction

$$C_nH_m + \frac{1}{2}\left(2n + \frac{m}{2}\right)O_2 \rightarrow nCO_2 + \frac{m}{2}H_2O.$$

CO_2 is taken to be a gas and water to be a liquid, in which case the heat of combustion corresponds to what is sometimes called the *higher heating value* of the fuel. Since all combustion reactions are exothermic, heats of combustion are tabulated without the negative sign, which is a most unfortunate convention. The computation is generally not necessary for hydrocarbons of fewer than ten carbons, where ΔH_F is generally tabulated separately.

EXAMPLE 15A.3 Compute the heat of formation of liquid n-octane, C_8H_{18}, from the tabulated heat of combustion, which is 5,470.7 kJ/mol.

$n = 8$ and $m/2 = 9$, so the heat of combustion is

$$\Delta \underline{H}_R = 8\Delta \underline{H}_{F,CO_2} + 9\Delta \underline{H}_{F,H_2O} - 12.5\Delta \underline{H}_{F,O_2} - \Delta \underline{H}_{F,C_8H_{18}}.$$

The heat of formation of the element oxygen is zero. The heats of formation of gaseous CO_2 and liquid H_2O are -393.5 and -285.8, respectively. We then have, taking care to include the negative sign with the heat of the combustion reaction,

$$-5470.7 = 8(-393.5) + 9(-285.8) - 12.5(0) - \Delta \underline{H}_{F,C_8H_{18}},$$

or $\Delta \underline{H}_{F,C_8H_{18}} = -249.5$ kJ/mol. To within rounding error, this is the same as the value for the heat of formation given in Table 15A.1.

Appendix 15B: Transient Behavior of an Adiabatic CFSTR

The graphical approach employed in Section 15.6 to examine the steady-state behavior of a CFSTR with a single reaction can be extended in a straightforward manner to understand the transient behavior for the special case of an adiabatic reactor; this restriction is necessary because the graphical approach permits analysis of only two differential equations. The mass balance and energy equations are

$$\theta \frac{dc_A}{dt} = c_{Af} - c_A - k_o\theta e^{-E/RT} c_A, \qquad (15B.1a)$$

$$\theta \frac{dT}{dt} = T_f - T + Jk_o\theta e^{-E/RT} c_A. \qquad (15B.1b)$$

These equations can be solved for c_A as follows:

Equation 15B.1a: $c_A = \dfrac{c_{Af}}{1 + k_o\theta e^{-E/RT}} - \dfrac{\theta \dfrac{dc_A}{dt}}{1 + k_o\theta e^{-E/RT}},$ $\qquad (15B.2a)$

Equation 15B.1b: $c_A = \dfrac{(T - T_f)}{Jk_o\theta e^{-E/RT}} + \dfrac{\theta \dfrac{dT}{dt}}{Jk_o\theta e^{-E/RT}}.$ $\qquad (15B.2b)$

When $dc_A/dt = 0$, Equation 15B.2a leads to Line I for c_A versus T in Figure 15.3, redrawn in Figure 15B.1. This line divides the $c_A - T$ plane into two regions: Above the line, where $c_A > c_{Af}/\left(1 + k_o\theta e^{-E/RT}\right)$, it follows from Equation 15B.2a that $dc_A/dt < 0$ and c_A is decreasing with time. Similarly, below the line, $dc_A/dt > 0$ and c_A is increasing with time.

We can do the same with Equation 15B.2b. When $dT/dt = 0$ we obtain Line III for c_A versus T, shown in Figure 15B.2. By differentiating c_A with respect to T in the steady-state equation and setting the derivative to zero, it is established that the maximum and minimum in Line III occur at $T_{\max,\min} = \dfrac{E \pm E^{1/2}(E - 4RT_f)^{1/2}}{2R}$, from which it follows that a maximum and minimum will occur at long as $E > 4RT_f$, which will

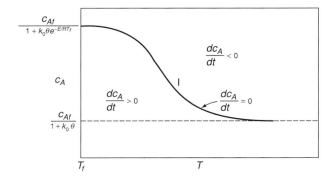

Figure 15B.1. Regions of positive and negative values of dc_A/dt in the $c_A - T$ plane. Line I, the locus of $dc_A/dt = 0$, is computed from Equation 15B.2a.

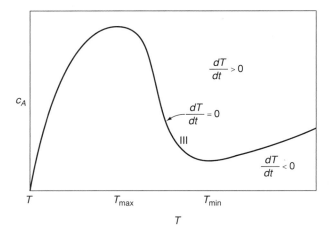

Figure 15B.2. Regions of positive and negative values of dT/dt in the $c_A - T$ plane. Line III, the locus of $dT/dt = 0$, is computed from Equation 15B.2b.

generally be the case. When c_A lies above Line III it follows from Equation 15B.2b that $dT/dt > 0$ and T increases with time. For c_A below the line, $dT/dt < 0$ and T decreases with time.

Figure 15B.3 shows Lines I and III superimposed. The curves are shown intersecting three times, but note that by adjusting the J, $k_o\theta$, and T_f they could be shifted relative to one another so that only one intersection is possible. Since these lines represent $dc_A/dt = 0$ and $dT/dt = 0$, their intersections correspond to steady-state

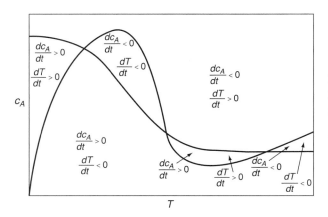

Figure 15B.3. Superposition of Figs. 15B.1 and 15B.2, showing regions of positive and negative dc_A/dt and dT/dt in the $c_A - T$ plane.

solutions of the coupled equations. The two lines divide the $c_A - T$ plane into six regions, and the algebraic signs of dc_A/dt and dT/dt are shown in each. The inequalities are replaced by arrows in Figure 15B.4; an arrow pointing to the left and up, for example, means $dT/dt < 0$ and $dc_A/dt > 0$, or "decreasing T, increasing c_A." By following the arrows we can determine how c_A and T will change as time progresses. Note that all arrows move away from the middle steady state, indicating its unavailability. By following the arrows we find that when we are to the right of Line IV, known as the *separatrix,* we will always go to the high-temperature steady state, whereas to the left we will always go to the low-temperature steady state. Thus, we know what conditions must prevail when starting the reactor in order to reach the desired steady-state operating condition. This is a particularly simple way of determining the evolution of the temperature-composition behavior, but it gives no information about the time required for the process to occur. Time dependence can only be obtained by solving the differential Equations 15B.1a,b.

This method of solution can be formalized by noting that the slopes of the arrows represent the rate of change of c_A with respect to T at any point in the $c_A - T$ plane, which is sometimes known as a *phase plane*; we write

$$\frac{dc_A}{dT} = \frac{dc_A/dt}{dT/dt} = \frac{c_{Af} - c_A - k_o\theta e^{-E/RT}c_A}{T_f - T + Jk_o e^{-E/Rt}c_A}.\qquad(15B.3)$$

Thus, for any pair (c_A, T) we can compute the slope. This is known as the *method of isoclines*. The method of isoclines is a powerful way to get information about the dynamics of nonlinear systems, but it is restricted to systems that can be described by a pair of differential equations.

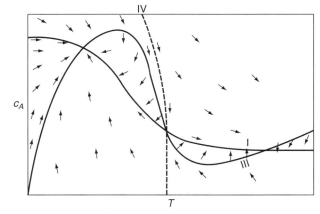

Figure 15B.4. Directions of change of c_A and T computed from the algebraic signs of dc_A/dt and dT/dt.

Postface

As noted in the first chapter, chemical engineering is a broad profession that is critical to addressing many of the issues facing modern society. The intent of this text has been to provide a fundamental understanding of the elements of chemical engineering and to provide a flavor of the challenges that a chemical engineer might face; the quantitative skills developed here are generalizable to problems of far greater complexity than those addressed in this introductory text. There is much more to come to complete a basic chemical engineering education; the core will normally include courses that cover thermodynamics, fluid mechanics, mass transfer, heat transfer, separations, and reactor analysis in depth, as well as a capstone course in design. Other courses in the curriculum will depend on the institution, but will include some selection of advanced courses in chemistry, materials, biology, and mathematics.

Most educational institutions offer undergraduate students an opportunity to do research, and this experience is invaluable for obtaining real insight into the scope of the profession – it is a truism that the research that chemical engineers do is rarely reflected in the courses in the undergraduate curriculum because of time limitations in a four-year professional program, and it is in the research laboratory that an undergraduate student is most likely to see the exciting topics in materials development, synthetic biology, nanotechnology, and so forth that were mentioned in Chapter 1, as well as to experience the intellectual excitement that comes with addressing real open-ended problems. Whatever your ultimate professional goal – professional practice with a bachelor's or master's degree, a PhD and a career in research or education, or further study in a related or even unrelated discipline – the tools developed here will serve you well throughout your career.

Index